Fate of Free, "Masked" and Conjugated/Modified forms of Mycotoxins

Fate of Free, "Masked" and Conjugated/Modified forms of Mycotoxins

Editor

Michele Suman

MDPI • Basel • Beijing • Wuhan • Barcelona • Belgrade • Manchester • Tokyo • Cluj • Tianjin

Editor
Michele Suman
Barilla G. R. F.lli SpA,
Advanced Laboratory Research
Italy

Editorial Office
MDPI
St. Alban-Anlage 66
4052 Basel, Switzerland

This is a reprint of articles from the Special Issue published online in the open access journal *Toxins* (ISSN 2072-6651) (available at: https://www.mdpi.com/journal/toxins/special_issues/Masked_Conjugated_Mycotoxins).

For citation purposes, cite each article independently as indicated on the article page online and as indicated below:

LastName, A.A.; LastName, B.B.; LastName, C.C. Article Title. *Journal Name* **Year**, *Volume Number*, Page Range.

ISBN 978-3-0365-0158-1 (Hbk)
ISBN 978-3-0365-0159-8 (PDF)

Cover image courtesy of Michele Suman.

Contents

About the Editor

Michele Suman was born in Rovigo, Italy, in 1973. He is married with two children. Michele graduated with a degree in Analytical Chemistry, Summa Cum Laude, from the University of Ferrara in 1997. He won the National Prize for Young Researchers promoted by the Italian Chemistry Federation (Federchimica) in 1998. He then completed a master's degree in 1998 (Master in Science, Technology and Management from University of Ferrara), while simultaneously working at the "Natta Research Center" of Shell-Montell Polyolefins, and a doctorate in 2005 (Ph.D. in Science and Technology of Innovative Materials from University of Parma) He landed in the role of Food Safety & Authenticity Research Manager of the Barilla Spa company in 2003. Here he has been working in an international context with public and private research centers organizations on research projects within the field of food chemistry, food safety-quality-authenticity, food contact materials, sensing and mass spectrometry applications for food products. He is Chair of the Italian National Normative Organization (UNI) Food Authenticity Working Group, a member of working groups for Biotoxins-Processing Contaminants-Food Authenticity in the European Committee for Standardization (CEN), Chair of the ILSI Process Related Compounds & Natural Toxins Task Force, a member of the Scientific Committee of Italian National Cluster Agrifood, a member of the Board of Mass Spectrometry Division – Italian Chemistry Society from 2014 to 2019, a member of the Scientific Board of Italian Chemistry Society—Food Chemistry Inter-divisional Group, and on the editorial boards of important peer-reviewed journals (*World Mycotoxin Journal*, *Food Additives and Contaminants*, etc.). He has been involved in various National/European funded projects and, presently, he is the WP Leader of the EU-FP7 FoodIntegrity Project and part of the EU-H2020 MyToolBox projects. He has experience in academic teaching activities, masters Ph.D. projects supervision, coordination/chairmanship of international conferences (Recent Advances in Food Analysis, World, Mycotoxins Forum, FoodIntegrity, MS Food Day, Rapid Methods Europe, and International Mass Spectrometry Conference). He is Adjunct Professor of AgriFood Authenticity at Catholic University of the Sacred Heart – Milan/Piacenza (Italy). His scientific production is documented by 7 books (editor/chapters), more than 150 contributions at national and international conferences, and more than 90 papers in international ISI journals.

 **_toxins_**

Editorial

Fate of Free and Modified Forms of Mycotoxins during Food Processing

Michele Suman

Barilla G. R. F.lli SpA, Advanced Research Labs, via Mantova 166, 43122 Parma, Italy;
michele.suman@barilla.com

Received: 22 May 2020; Accepted: 2 July 2020; Published: 10 July 2020

International trade is highly affected by mycotoxin contaminations, which result in an annual 5% to 10% loss of global crop production [1]. In the last decade, the mycotoxin scenario has been complicated through the progressive understanding—beside emerging mycotoxins—of the parallel presence of modified (masked and conjugated) forms, in addition to the previously free known ones.

The present *Toxins* Special Issue provides original research papers and reviews that deal with the fates of all these forms of mycotoxins, with respect to aspects that cover traditional and industrial food processing, yearly grain campaign peculiar conditions and management, novel analytical solutions, consumer exposure, and biomarkers-assessment directions.

Among emerging mycotoxins, *Fusarium* ones, such as fusaproliferin (FUS), beauvericin (BEA), enniatins (ENNs), and moniliformin (MON) are discussed within the Serbian maize context by Jajić et al., highlighting exactly the economic impact of these mycotoxins in terms of the yield and quality of grain along different yearly campaigns [2]. MON, BEA and FUS are indicated as being major contaminants in more than half of the analyzed samples, and are considered in strict connection with climate change consequences which surely must be taken into account more and more in the future.

Mechanical & thermal energy involved in food processing determine changes into mycotoxin forms and the creating or destroying of new bonds with other food components: a clear example is reported by Zhang et al., describing the conversion of deoxynivalenol-3-glucoside (DON-3G) to deoxynivalenol (DON) during Chinese steamed bread processing, along the fermentation and steaming steps. Mechanical friction and shear seem to play roles which lead to these mycotoxins' structural changes, but only in combination with other parallel factors probably related to ingredients and complex physico-chemical modifications that occur and need further investigation [3].

In a global scenario, Schaarschmidt and Fauhl-Hassek consider South American atmospheres and traditions with their review about mycotoxins' changes during the processes of nixtamalization and tortilla production [4]. Alkaline cooking has been proven effective for reducing aflatoxins and fumonisins in cooked maize and tortillas, even if acidic conditions could partially reverse this process.

These phenomena must be deeply understood in the future for assuring that the benefits concerning the formation of low toxic hydrolyzed fumonisins are not negatively balanced out by the parallel formation of other toxicologically relevant modified and matrix-associated forms.

Remaining in the context of bakery products, can we properly design and optimize industrial baking conditions to mitigate processing contaminants and mycotoxins, while not heavily affecting the organoleptic aspects, in one single shot? The answer of Suman et al. is yes! This answer is corroborated by scientific evidence on how acrylamide concentration may be influenced by wholegrain and cocoa biscuit bakery-making parameters within a parallel strategy of DON mitigation, highlighting a significant role of pH, followed by the baking time/temperature parameters [5].

Stadler et al. focus their attention on bakery production and in particular on the optimization of recipes and processing parameters at an industrial scale, devoted to the mitigation of the main mycotoxin contaminant in the common wheat chain: DON [6]. DON degradation is accurately quantified in industrially made crackers, biscuits, and bread, showing how degradation (setting properly raising agents and baking times/temperatures) means, practically, conversion into a less toxic isomeric product (isoDON), with correspondingly positive implications towards the safety of the final consumer with regard to these commodities, with respect to the original contamination of the exploited raw material.

Moving from bakery production to meal solutions, Tittlemier et al. show changes into ergot alkaloids pattern along the durum wheat pasta production chain. More than 80% of the total ergot alkaloids are confined into outer kernel layers after milling; ergocristine, ergocristinine, and ergotamine remain the predominant components which do not also decrease after pasta production and cooking steps. Besides, the milling and cooking of pasta alters the ratio of R- to S-enantiomers; this epimerization results in a higher final concentration of the less biologically active S-enantiomers in boiled spaghetti [7].

Sueck et al. rightly consider some doubts regarding the real effect of thermal processing in food commodities such as coffee or bread; this does not always imply a degrading/detoxifying action, but, in some instances, determines the formation of unexpected forms, the toxicological definitive evaluations that would permit an adequate overall risk assessment for which are still missing. In their recent study, they specifically demonstrate the generation of the isomerization product, 20R-ochratoxin A (20R-OTA), from ochratoxin A (OTA) [8].

Ksieniewicz-Wozniak et al. are keen on the consumers' exposure scenario, looking into the beer production chain, demonstrating that, within *Fusarium* mycotoxins, DON and its main metabolite DON-3G, among the samples analyzed, are present practically everywhere [9]. Then, nivalenol (NIV) and nivalenol-3-glucoside (NIV-3G) were also found to be largely present in both malt samples and beers. Their conclusion sounds like a warning: *Fusarium* mycotoxins should not be overlooked in countries with a very high beer consumption. In the worst-case scenario the probable daily intake (PDI) would exceed the tolerable daily intake (TDI) with only one half-liter bottle!

De Santis et al. provide relevant findings about the useful exploitation of urinary DON and its glucuronide and de-epoxydated (DOM-1) forms as biomarkers for exposure assessment purposes, permitting us to identify particularly vulnerable categories, such as children and adolescent age groups [10].

Finally, there is the evident necessity for all the stakeholders (from authorities, to control bodies and food business operators) to dispose of rapid and easy-to-use methods for the determination of free and modified forms of toxins in raw materials; Lippolis et al. propose here a fluorescence polarization immunoassay (FPIA) for the simultaneous determination of T-2 toxin, HT-2 toxin and relevant glucosides, expressed as sum, exploiting a HT-2-specific antibody with high sensitivity and high cross-reactivity towards the different forms present in cereals [11]. This analytical method is compliant with harmonized guidelines for the validation of screening methods recently stated by European regulations.

I would like to congratulate and thank all the authors involved in this special issue of *Toxins* and I hope you enjoy reading its contents; it gives a taste of an exciting scientific field which has several implications into our daily life, because (i) it covers our diet practically and from every point of view, (ii) it intersects our culinary uses and customs, but also industrial production processes, and (iii) it involves a careful evaluation of costs and benefits and a constant continuous improvement of mitigation strategies. There will still be a lot to see and discover in the coming years!

Funding: This research received no external funding.

Conflicts of Interest: The author declares no conflict of interest.

References

1.   EC. European Commission Decision C 2453: Horizon 2020 Work Programme 2014–2015: Food Security, Sustainable Agriculture and Forestry, Marine and Maritime and Inland Water Research and Bioeconomy (Revised). 2015. Available online: https://ec.europa.eu/research/participants/data/ref/h2020/wp/2014_2015/ main/h2020-wp1415-food_en.pdf (accessed on 7 June 2020).

2. Jajić, I.; Dudaš, T.; Krstović, S.; Krska, R.; Sulyok, M.; Bagi, F.; Savić, Z.; Guljaš, D.; Stankov, A. Emerging Fusarium Mycotoxins Fusaproliferin, Beauvericin, Enniatins, and Moniliformin in Serbian Maize. *Toxins* **2019**, *11*, 357. [CrossRef]

3. Zhang, H.; Wu, L.; Li, W.; Zhang, Y.; Li, J.; Hu, X.; Sun, L.; Du, W.; Wang, B. Conversion of Deoxynivalenol-3-Glucoside to Deoxynivalenol during Chinese Steamed Bread Processing. *Toxins* **2020**, *12*, 225. [CrossRef] [PubMed]

4. Schaarschmidt, S.; Fauhl-Hassek, C. Mycotoxins during the Processes of Nixtamalization and Tortilla Production. *Toxins* **2019**, *11*, 227. [CrossRef] [PubMed]

5. Suman, M.; Generotti, S.; Cirlini, M.; Dall'Asta, C. Acrylamide Reduction Strategy in Combination with Deoxynivalenol Mitigation in Industrial Biscuits Production. *Toxins* **2019**, *11*, 499. [CrossRef] [PubMed]

6. Stadler, D.; Lambertini, F.; Woelflingseder, L.; Schwartz-Zimmermann, H.; Marko, D.; Suman, M.; Berthiller, F.; Krska, R. The Influence of Processing Parameters on the Mitigation of Deoxynivalenol during Industrial Baking. *Toxins* **2019**, *11*, 317. [CrossRef] [PubMed]

7. Tittlemier, S.A.; Drul, D.; Roscoe, M.; Turnock, D.; Taylor, D.; Fu, B.X. Fate of Ergot Alkaloids during Laboratory Scale Durum Processing and Pasta Production. *Toxins* **2019**, *11*, 195. [CrossRef] [PubMed]

8. Sueck, F.; Hemp, V.; Specht, J.; Torres, O.; Cramer, B.; Humpf, H.-U. Occurrence of the Ochratoxin A Degradation Product 2′R-Ochratoxin A in Coffee and Other Food: An Update. *Toxins* **2019**, *11*, 329. [CrossRef] [PubMed]

9. Ksieniewicz-Woźniak, E.; Bryła, M.; Waśkiewicz, A.; Yoshinari, T.; Szymczyk, K. Selected Trichothecenes in Barley Malt and Beer from Poland and an Assessment of Dietary Risks Associated with their Consumption. *Toxins* **2019**, *11*, 715. [CrossRef] [PubMed]

10. De Santis, B.; Debegnach, F.; Miano, B.; Moretti, G.; Sonego, E.; Chiaretti, A.; Buonsenso, D.; Brera, C. Determination of Deoxynivalenol Biomarkers in Italian Urine Samples. *Toxins* **2019**, *11*, 441. [CrossRef] [PubMed]

11. Lippolis, V.; Porricelli, A.C.R.; Mancini, E.; Ciasca, B.; Lattanzio, V.M.T.; De Girolamo, A.; Maragos, C.M.; McCormick, S.; Li, P.; Logrieco, A.F.; et al. Fluorescence Polarization Immunoassay for the Determination of T-2 and HT-2 Toxins and Their Glucosides in Wheat. *Toxins* **2019**, *11*, 380. [CrossRef] [PubMed]

Article

Conversion of Deoxynivalenol-3-Glucoside to Deoxynivalenol during Chinese Steamed Bread Processing

Huijie Zhang [1,2], **Li Wu** [1,2], **Weixi Li** [1,2], **Yan Zhang** [1,2], **Jingmei Li** [1,2], **Xuexu Hu** [1,2], **Lijuan Sun** [1,2], **Wenming Du** [1,2] and **Bujun Wang** [1,2,*]

[1] Institute of Crop Science, Chinese Academy of Agricultural Sciences, Beijing 100081, China; zhanghuijie01@caas.cn (H.Z.); wuli02@caas.cn (L.W.); liweixi@caas.cn (W.L.); zhangyan06@caas.cn (Y.Z.); lijingmei@caas.cn (J.L.); huxuexu@caas.cn (X.H.); sunlijuan@caas.cn (L.S.); duwenming@caas.cn (W.D.)

[2] Laboratory of Quality & Safety Risk Assessment for Cereal Products (Beijing), Ministry of Agriculture, Beijing 100081, China

* Correspondence: wangbujun@caas.cn; Tel.: +86-010-8210-5798

Received: 2 March 2020; Accepted: 30 March 2020; Published: 3 April 2020

Abstract: We reported the conversion of deoxynivalenol-3-glucoside (D3G) to deoxynivalenol (DON) during Chinese steamed bread (CSB) processing by artificial D3G contamination. Meanwhile, the effects of enzymes in wheat flour and those produced from yeast, along with the two main components in wheat flour—wheat starch and wheat gluten—on the conversion profiles of D3G were evaluated. The results showed D3G could convert to DON during CSB processing, and the conversion began with dough making and decreased slightly after fermentation and steaming. However, there was no significant difference in three stages. When yeast was not added, or enzyme-deactivated wheat flour was used to simulate CSB process, and whether yeast was added or not, D3G conversion could be observed, and the conversion was significantly higher after dough making. Likewise, D3G converted to DON when wheat starch and wheat gluten were processed to CSB, and the conversion in wheat starch was higher.

Keywords: deoxynivalenol-3-glucoside; deoxynivalenol; conversion; Chinese steamed bread; processing

Key Contribution: D3G could convert to DON during Chinese steamed bread (CSB) processing, the conversion started from dough making, and this conversion could also occur when wheat starch and wheat gluten were processed to CSB. In addition, D3G could convert to DON without yeast or enzymes in wheat flour or both of them.

---

1. Introduction

Trichothecenes are the most prevalent mycotoxin family, which are mainly produced by *Fusarium* species such as *F. culmorum* and *F. graminearumand*. Deoxynivalenol (DON), in particular, is the best known and predominant one due to its worldwide occurrence in cereals and products derived from them [1]. Due to the importance of wheat in diet, it is of concern that *Fusarium* can infect grains and produce mycotoxins under certain climate conditions, which can contaminate final food products. In recent years, interest has been growing in toxins called masked mycotoxins that are primarily produced in plants by enzymatic transformations related to plant resistance mechanisms to counteract pathogen invasions [2–6]. Masked mycotoxins, usually formed by the reaction of parent mycotoxins with amino acids or sugars, therefore, occur in conjugated forms. Furthermore, the modification may occur in the food matrix, by covalent binding or non-covalent association to sugars, proteins, or other macromolecules [7]. Deoxynivalenol-3-glucoside (D3G) is the most known masked mycotoxin, which is produced by conjugation of DON and glucose, and has been detected in various foods, such as breakfast cereals, bread, and beers [8–10]. The main concern about masked mycotoxins is that their conjugates can be hydrolyzed in the stomachs of mammals after uptake, thereby releasing the toxic precursor DON and influencing its bioaccessibility [11]. The Joint FAO/WHO Expert Committee on Food Additives (JECFA) declared that D3G could contribute to DON dietary exposure; research on its absorption, distribution, metabolism, excretion in animal and human body, and its fate in food processing are needed [12]. The acetylated derivatives of DON, 15-acetyl-deoxynivalenol (15-AcDON) and 3-acetyl-deoxynivalenol (3-AcDON), are intermediate products of fungal DON biosynthesis that generally occur together with DON in cereal commodities, and 3-AcDON can convert to DON during mammalian metabolic processes and thus contribute to the total DON toxicity [13]. Therefore, AcDONs are considered to be masked mycotoxins by some researchers [7], and the JECFA amended the provisional maximum tolerable daily intake (PMTDI) for DON to 1 mg/kg bodyweight for DON and its acetylated forms [12].

DON's glucoside form has been found in various raw cereals. It was reported that the mean concentrations of D3G were 393 µg/kg and 141 µg/kg, respectively, for wheat and maize samples [14]. Palacios et al. (2017) [15] reported that D3G was detected in 94% of the investigated durum wheat commercial cultivars from the Argentinean main growing area at concentrations ranging from < the limit of quantification (LOQ, 50 µg/kg) to 850 µg/kg. Lancova et al. (2008) [16] found D3G in naturally contaminated barley, and a remarkable increase of D3G was observed in malt compared to raw barley. Ksieniewicz-Wo'zniak et al. (2019) [17] investigated D3G levels in 87 barley malt samples, and 91% were positive, with concentrations ranging from 4.4–410.3 µg/kg. Quite a lot of similar reports have been presented [18–20].

Previous research indicates that mycotoxins may transform in food processing via heating, fermentation, or from ingredients such as enzymes [21]. It was also reported that food processes, such as sorting, cleaning, milling, brewing, baking, frying, roasting, alkaline cooking, extrusion, etc., might affect mycotoxins [22]. Many agricultural products that contaminated with mycotoxins are processed using germination (barley, for example), fermentation, hydrolysis, enzymes, and alkaline or acidic hydrolytic conditions, which contribute to the production or release of masked mycotoxins [23]. D3G can be transformed to DON in food processing, and vice versa [24]. The cleavage of masked mycotoxins occurs during the process of malting, leading to an increase in free DON [8]. DON contents increased after baking, which suggests that bound DON is released in baking [25].

Among the many factors influencing masked mycotoxin transformation during food processing, enzymes produced from microorganisms have garnered substantial interest from researchers worldwide. It was reported that doughnuts fermented with yeast contaminated higher DON content than that in the flour, and this may be due to enzymatic transformation [26]. Simsek et al. (2012) [25] reported the effects of milling and baking processing on D3G. The addition of enzyme mixtures to improve baking increased the content of D3G in fermented dough up to 145%. D3G and DON levels decreased slightly in baking. When whole wheat was treated with enzymes to evaluate the effects of enzymatic hydrolysis on DON, the results showed that DON contents were prominently higher after treatment with protease (16%) and xylanase (39%), which suggested DON maybe embed or bind in the cell wall or protein component of wheat kernel. Bread processing usually involves fermentation with yeast or leavening agents (leaven). In particular, rye bread with leaven requires more extensive enzymatic activity [27].

In our previous study, we found that the DON levels in baked bread were almost double those in flour, while the D3G contents in baked bread were notably lower. The increase in DON and the decrease in D3G started during the fermentation of the dough. DON contents approximately doubled after mixed and fermented dough was processed to CSB, and D3G concentrations were almost 50% lower than in flour, suggesting that CSB processing may release bound DON in the flour [28,29]. Our research team investigated the fate of 3-ADON and 15-ADON in bread processing by spiking mycotoxin-free wheat flour with 3-ADON and 15-ADON standards, and the results showed that ADONs could convert to DON during bread processing [30].

The objective of the assay presented in this study was to investigate whether D3G could convert to DON during CSB processing, as suggested by our previous research. The effects of enzymatic hydrolysis and different wheat compositions on D3G conversion were verified.

2. Results

2.1. Conversion of D3G during CSB Processing

In our previous study, wheat flour naturally contaminated with DON and D3G was used for wheat-based product processing. Thus, it was generally contaminated with small amounts of other derivatives of DON, especially other masked DONs. To elucidate the behavior of D3G during Chinese steamed bread processing, D3G contaminated wheat flour samples were prepared in this study by spiking wheat flour, free of target mycotoxins, with a standard solution of D3G.

DON concentrations in the doughs (mixed dough and fermented dough) and steamed products produced from wheat flour spiked with three different D3G levels are presented in Table 1. DON was detected in the whole processing of CSB, and the amount of DON converted from D3G was proportional to the spiked amount. In addition, the DON concentrations converted from D3G during the whole processing decreased, and there was no significant difference ($p < 0.05$) among mixed dough, fermented dough, and steamed products. The results indicated that D3G could release DON in CSB processing and that the conversion started with the dough making process. The results agreed with those from our previous research, in which D3G concentration changes started with the dough making process [28].

Table 1. Deoxynivalenol (DON) concentrations converted from spiked deoxynivalenol-3-glucoside (D3G) at different stages of Chinese steamed bread processing.

Samples	Spiking Levels (µg/kg)		
	300	500	800
Mixed dough	26.02 ± 2.07a	34.64 ± 5.42a	61.75 ± 4.88a
Fermented dough	26.37 ± 2.11a	31.55 ± 4.96a	54.42 ± 5.16a
Chinese steamed bread	21.33 ± 3.37a	25.96 ± 3.46a	52.16 ± 5.69a

DON concentrations were average values based on three replicates. Means followed by the same small letters within columns are not significantly different ($p > 0.05$).

2.2. Role of Enzymes in the Conversion of D3G during CSB Processing

Previous research suggests that D3G can release DON during food processing as a result of enzymatic degradation of polysaccharides [25,26]. Yeast and wheat flour were two sources of enzymes in CSB processing. Wheat flour contains several important enzymes, such as amylases, proteases, lipoxygenase, polyphenol oxidase, and peroxidase. These enzymes are inactive during grain and flour storage, and they become active when water is added [31]. Enzymes produced by yeast also play an important role in dough fermentation. The results of the present study showed that D3G could convert to DON during CSB processing without using yeast (Treatment 1) and when using enzyme-deactivated wheat flour with the addition of yeast (Treatment 2). The conversion started from dough making, as observed in the CSB processing of wheat flour with added yeast (CK) (Figure 1). In addition, the conversion was also observed when enzyme-deactivated wheat flour was processed to CSB, and with no yeast used (Treatment 3). While DON concentrations converted from D3G after dough making in these three treatments were significantly higher than that of the control, they decreased dramatically after dough fermentation and decreased or increased slightly after steaming. Except when enzyme-deactivated wheat flour with no yeast was used in CSB processing, the DON levels in fermented dough and steamed products were significantly lower than those of the control. These results suggested that D3G could convert to DON during CSB processing in the absence of yeast or enzymes in wheat flour or both of them, while the production of DON in steamed products was generally lower. The significant increase in the concentration of DON after dough making merits further research.

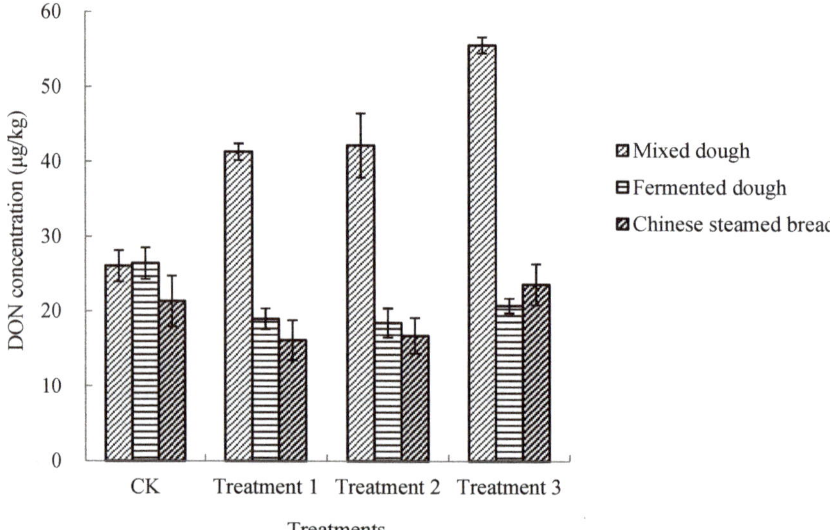

Figure 1. DON concentrations converted from spiked D3G (300 µg/kg) in different treatments of Chinese steamed bread. CK: wheat flour, yeast used; Treatment 1: wheat flour, no yeast used; Treatment 2: enzyme-deactivated wheat flour, yeast used; Treatment 3: enzyme-deactivated wheat flour, no yeast used.

2.3. Conversion of D3G in Different Wheat Compositions during CSB Processing

The fate of mycotoxins in food processing was affected by several factors, such as the food matrix, pH, moisture content, temperature, natural or spiked contamination, and original mycotoxin concentration [32]. To determine whether the compositions in flour play an important role in D3G conversion during CSB processing, wheat starch and wheat gluten were chosen to simulate CSB processing, since they account for approximately 75% and 10%, respectively, of wheat flour. The results showed that D3G conversion also occurred during the CSB processing of wheat starch and wheat gluten (Figure 2). The conversion of D3G in wheat starch was significantly higher than that of wheat flour and wheat gluten in dough mixing, fermentation, or steaming, and the conversion amounts of D3G in wheat gluten was the lowest. As observed in the CSB processing of three wheat flour samples spiked with three different D3G levels, the DON conversion from D3G during the CSB processing of wheat starch and wheat gluten decreased. No data are available on the conversion of D3G in different wheat compositions during CSB processing. The results of this study suggest that amylose or amylopectin in wheat flour may play a crucial role in D3G conversion during CSB processing, and that this mechanism merits further study.

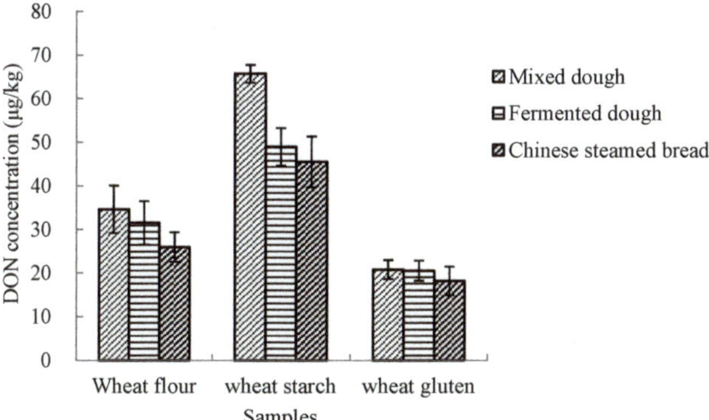

Figure 2. DON concentrations converted from spiked D3G (500 µg/kg) during Chinese steamed bread processing with different wheat compositions.

3. Discussion

This study confirmed that D3G could convert to DON during CSB processing, and the conversion started from the dough making process and decreased slightly after dough fermentation and steaming. In addition, D3G could convert to DON during CSB processing in the absence of yeast or enzymes in wheat flour or both of these components, while converted DON in steamed products was generally lower. In addition, D3G conversion could occur when wheat starch and wheat gluten were processed to CSB, and the amounts converted in wheat starch were significantly higher.

The results of this work indicated that the conversion of D3G in CSB processing began with dough mixing. Our previous research indicated that DON levels increased significantly after dough was rolled in a noodle machine during noodle processing, and that D3G levels decreased dramatically [29]. These studies suggested that mechanical force may exert a remarkable influence on mycotoxin structure, therefore leading to the liberation of masked mycotoxins. To verify this hypothesis, we studied the effect of mechanical forces (oscillate in an ultrasonic cleaner and blend at high speed in a blender for 30 min, respectively) on the conversion of D3G by spiking D3G standard in water and acetonitrile. The results (data not listed) showed that only trace amounts of D3G (approximately 1%) converted to DON after oscillating or blending. Further research indicated 15-AcDON and 3-AcDON could also convert to their parent mycotoxins when 15-AcDON and 3-AcDON standard spiked in water or acetonitrile were oscillated in an ultrasonic cleaner, blended at high speed in a blender, shaken in a shaking incubator, or stirred in a magnetic stirring stirrer; likewise, only a little bit of DON could be detected after treating. The results suggested that mechanical force had an impact on the structures of D3G, 15-AcDON, and 3-AcDON, leading to the conversion to DON, and the conversion of D3G standard (in acetonitrile) under mechanical forces indicated that the hydroxyl of water molecules was not a crucial factor in D3G conversion. There is no description of this phenomenon in the existing literature. This may lead us to a new realm called mechanochemistry, which involves mechanical and chemical behaviors on a molecular level, and various phenomena, such as mechanical breakage, polymeride degradation by shearing, cavitation-involved phenomena, shock wave chemistry and physics, molecular machines, etc., are included. Mechanochemistry could be regarded as the intersection of mechanical engineering and chemistry, and it produces chemical product synthesis depending on possible mechanical actions. The mechanisms of mechanochemical transformations are different from those of general techniques [33]. Ball milling is a widely used technology, in which chemical processing and transformations are moved by mechanical force [34]. In this study, the generation of DON under mechanical forces may occur due to the mechanical energy of mechanical friction and shear exerted upon mycotoxin molecules that lead to structural changes of D3G, 15-AcDON, and 3-AcDON. However, as only a trace amount of D3G could convert to DON in liquid under mechanical forces, and CSB processing generally involves different compositions in a flour matrix, ingredients and complex physico-chemical modifications that occur in the processing, D3G conversion in CSB processing may be due to other mechanisms, and it still needs further research.

4. Materials and Methods

4.1. Chemicals and Reagents

The analytical standard of D3G (50 μg/mL in acetonitrile, certified purity >99.9%) was purchased from Sigma (Sigma-Aldrich, Alcobendas, Spain). Purified water was obtained from a Milli-Q apparatus (Millipore Corp., Bedford, MA, USA). Methanol, acetonitrile and formic acid (all HPLC grade) were purchased from Thermo Fisher Scientific Corporation (Shanghai, China).

4.2. Preparation of D3G-Contaminated Wheat Flour/Enzyme-Deactivated Wheat Flour/Wheat Starch/Wheat Gluten

In the present experiment, wheat flour, wheat starch and wheat gluten samples with undetected levels of target mycotoxins were designated to be "blank". Wheat flour was purchased from Huanghua Jinmai Flour Co., Ltd. (Cangzhou, Hebei, China). Wheat starch was purchased from Shanghai Saiwengfu Agricultural Development Co., Ltd., and wheat gluten was from Shandong Qufeng Food Tech Co., Ltd. Blank wheat flour was dried at 130 °C in a drying oven (DHG-9140A, Shanghai Yiheng Scientific Instruments Co., Ltd., Shanghai, China) for three hours to prepare enzyme-deactivated blank wheat flour. D3G-contaminated wheat flour/enzyme-deactivated wheat flour/wheat starch/wheat gluten samples were prepared by spiking blank samples with a standard solution of D3G.

4.3. Preparation of Chinese Steamed Bread

Chinese steamed bread was processed according to the Chinese Business Standard (procedure 10139-93, Appendix A, 1993) with some modifications. The experiment was conceived and executed as follows: (1) blank wheat flour spiked with 3 levels of 300, 500, and 800 μg/kg D3G (based on the dry sample as delineated below), mixed with yeast solution (0.26 g dry yeast dispersed in 12.5 mL water at 38 °C, as delineated below); (2) blank wheat flour spiked with 300 μg/kg D3G, mixed with 12.5 mL water (38 °C) instead of yeast solution; (3) enzyme-deactivated blank wheat flour spiked with 300 μg/kg D3G, mixed with 12.5 mL yeast solution, and an additional 12.5 mL water (38 °C) was added to form the resultant dough; (4) enzyme-deactivated blank wheat flour spiked with 300 μg/kg D3G, mixed with 25 mL water (38 °C) instead of yeast solution; (5) blank wheat starch spiked with 500 μg/kg D3G, mixed with 12.5 mL yeast solution; (6) blank wheat gluten spiked with 500 μg/kg D3G, mixed with 12.5 mL yeast solution and additional 7.5 mL water (38 °C). Each treatment was prepared in triplicate. After 3 min of mixing, the mixed doughs were fermented in a fermentation cabinet (38 °C, 85% RH) for 60 min. Then manually molding to produce a round dough with a smooth surface was performed for 3 min, followed by steaming at 100 °C in a steaming chamber for 20 min after putting in air for 15 min. Then the steamed dough was cooled at room temperature for 40–60 min. Sampling was conducted at the end of dough preparation, after fermentation, and after steaming. Representative subsamples were stored at −20 °C until analysis.

4.4. Sample Treatment and UPLC-MS/MS Analysis

Sample treatment was performed as described by Zhang and Wang (2014) [28] with slight modifications. Ten mL acetonitrile: water (80:20, *v/v*) was blended with two grams of a representative sample and extracted for 3 min with a blender (IKA Co., Staufen, Germany). Then the sample was centrifuged at 10,000 rpm for 10 min, and the supernatant (5 mL) was filtered through a multifunctional MycoSep 226 columns (Romer Labs, Inc. Union, MO, USA), and 2 mL of the extract was submitted to N-EVAP at 50 °C until dry. Sequentially, the residue was dissolved in a mixture of 0.4 mL methanol: water (50:50, *v/v*), then vortexed and filtered through 0.22 μm MICRO PES filter (Membrana, Germany) for UPLC-MS/MS analysis.

The UPLC-MS/MS analysis of DON was conducted as described by Zhang and Wang (2014) [28] with some modifications. A multiple reaction monitoring (MRM) mode was performed, and column temperature was set at 26 °C, capillary voltage at 2.5 KV, cone voltage at 20 V. Gaseous nitrogen was used as desolvation gas, and its flows were maintained at 800 L/h. Desolvation temperature was set at 450 °C. Mobile phase A was methanol and mobile phase B was 0.1% (*v/v*) formic acid in water. Gradient of phase A performed was 0.2 mL/min for 0–3.5 min, and increased linearly from 5% to 85%, for 3.5–4.5 min a linear increase from 85% to 100% was followed, then decreased from 100% to 5% for 4.5–5.0 min, and followed by an isocratic washout of 5% A for 1 min.

Author Contributions: All authors conceived the experimental design. H.Z. performed the experiment, data statistical analysis, and wrote the original draft. L.W., W.L., and Y.Z. helped with laboratory analysis. J.L., X.H., L.S., and W.D. carried out the literature investigation. B.W. reviewed and edited the manuscript. All authors have read and agreed to the published version of the manuscript.

Funding: This research was funded by the National Key Research and Development Program of China (2016YFF0201803).

Conflicts of Interest: The authors declare no conflict of interest.

References

1. Pestka, J.J. Deoxynivalenol: Mechanisms of action, human exposure, and toxicological relevance. *Arch. Toxicol.* **2010**, *137*, 283–298. [CrossRef]
2. Engelhardt, G.; Ruhland, M.; Wallnöfer, P.R. Metabolism of mycotoxins in plants. *Adv. Food Sci.* **1999**, *21*, 71–78.
3. Karlovsky, P. Biological detoxification of fungal toxins and its use in plant breeding, feed and food production. *Nat. Toxins* **1999**, *7*, 1–23. [CrossRef]
4. Berthiller, F.; Werner, U.; Adam, G.; Krska, R.; Lemmens, M.; Sulyok, M.; Hauser, M.T.; Schuhmacher, R. Bildung von maskierten Fusarium mykotoxinen in Pflanzen. *Ernaehrung* **2006**, *30*, 477–481.
5. Lemmens, M.; Scholz, U.; Berthiller, F.; Dall'Asta, C.; Koutnik, A.; Schuhmacher, R.; Adam, G.; Buerstmayr, H.; Mesterházy, A.; Krska, R.; et al. The ability to detoxify the mycotoxin deoxynivalenol colocalizes with a major quantitative trait locus for Fusarium head blight resistance in wheat. *Mol. Plant Microbe Interact.* **2005**, *18*, 1318–1324. [CrossRef] [PubMed]
6. Schweiger, W.; Boddu, J.; Shin, S.; Poppenberger, B.; Berthiller, F.; Lemmens, M.; Muehlbauer, G.J.; Adam, G. Validation of a candidate deoxynivalenol-inactivating UDP-glucosyltransferase from barley by heterologous expression in yeast. *Mol. Plant Microbe Interact.* **2010**, *23*, 977–986. [CrossRef] [PubMed]
7. Cirlini, M.; Dall'Asta, C.; Galaverna, G. Hyphenated chromatographic techniques for structural characterization and determination of masked mycotoxins. *J. Chromatogr. A.* **2012**, *1255*, 145–152. [CrossRef]
8. Kostelanska, M.; Hajslova, J.; Zachariasova, M.; Malachova, A.; Kalachova, K.; Poustka, J.; Fiala, J.; Scott, P.M.; Berthiller, F.; Krska, R. Occurrence of deoxynivalenol and its major conjugate, deoxynivalenol-3-glucoside, in beer and some brewing intermediates. *J. Agric. Food Chem.* **2009**, *57*, 3187–3194. [CrossRef]
9. Malachova, A.; Dzuman, Z.; Veprikova, Z.; Vaclavikova, M.; Zachariasova, M.; Hajslova, J. Deoxynivalenol, deoxynivalenol-3-glucoside, and enniatins: The major mycotoxins found in cereal-based products on the Czech market. *J. Agric. Food Chem.* **2011**, *59*, 12990–12997. [CrossRef]
10. Vendl, O.; Berthiller, F.; Crews, C.; Krska, R. Simultaneous determination of deoxynivalenol, zearalenone, and their major masked metabolites in cereal based food by LCMS/MS. *Anal. Bioanal. Chem.* **2009**, *395*, 1347–1354. [CrossRef]
11. González-Arias, C.A.; Marín, S.; Sanchis, V.; Ramos, A.J. Mycotoxin bioaccessibility/absorption using in vitro digestion model: A review. *World Mycotoxin J.* **2013**, *6*, 167–184. [CrossRef]
12. JECFA (Joint FAO/WHO Expert Committee on Food Additives). Evaluation of Certain Contaminants in Food: Seventy-Second Report of the Joint FAO/WHO Expert Committee on Food Additives. WHO Technical Report Series, 959. 2011. Available online: http://whqlibdoc.who.int/trs/WHO_TRS_959_eng.pdf (accessed on 25 January 2014).
13. Bretz, M.; Beyer, M.; Cramer, B.; Humpf, H.U. Synthesis of stable isotope labeled 3-acetyldeoxynivalenol. *Mol. Nutr. Food Res.* **2005**, *49*, 1151–1153. [CrossRef] [PubMed]

14. Berthiller, F.; Dall'asta, C.; Corradini, R.; Marchelli, R.; Sulyok, M.; Krska, R.; Adam, G.; Schuhmacher, R. Occurrence of deoxynivalenol and its 3-beta-D-glucoside in wheat and maize. *Food Addit. Contam.* **2009**, *26*, 507–511. [CrossRef] [PubMed]

15. Palacios, S.A.; Erazo, J.G.; Ciasca, B.; Lattanzio, V.M.T.; Reynoso, M.M.; Farnochi, M.C.; Torres, A.M. Occurrence of deoxynivalenol and deoxynivalenol-3-glucoside in durum wheat from Argentina. *Food Chem.* **2017**, *230*, 728–734. [CrossRef]

16. Lancova, K.; Hajslova, J.; Poustka, J.; Krplova, A.; Zachariasova, M.; Dostalek, P.; Sachambula, L. Transfer of Fusarium mycotoxins and 'masked' deoxynivalenol (deoxynivalenol-3-glucoside) from field barley through malt to beer. *Food Addit. Contam.* **2008**, *25*, 732–744. [CrossRef]

17. Ksieniewicz-Woźniak, E.; Bryła, M.; Waśkiewicz, A.; Yoshinari, T.; Szymczyk, K. Selected Trichothecenes in barley malt and beer from Poland and an assessment of dietary risks associated with their consumption. *Toxins* **2019**, *11*, 715.

18. Berthiller, F.; Dall'Asta, C.; Schuhmacher, R.; Lemmens, M.; Adam, G.; Krska, R. Masked mycotoxins: Determination of a deoxynivalenol glucoside in artificially and naturally contaminated wheat by liquid chromatography-tandem mass spectrometry. *J. Agric. Food Chem.* **2005**, *53*, 3421–3425. [CrossRef]

19. Sasanya, J.J.; Hall, C.; Wolf-Hall, C. Analysis of deoxynivalenol, masked deoxynivalenol, and Fusarium graminearum pigment in wheat samples, using liquid chromatography–UV–mass spectrometry. *J. Food Protect.* **2008**, *71*, 1205–1213. [CrossRef]

20. Li, F.Q.; Yu, C.C.; Shao, B.; Wang, W.; Yu, H.X. Natural occurrence of masked deoxynivalenol and multi-mycotoxins in cereals from China harvested in 2007 and 2008. *Chin. J. Prev. Med.* **2011**, *45*, 57–63.

21. Berthiller, F.; Schuhmacher, R.; Adam, G.; Krska, R. Formation, determination and significance of masked and other conjugated mycotoxins. *Anal. Bioanal. Chem.* **2009**, *395*, 1243–1252. [CrossRef]

22. Bullerman, L.B.; Bianchini, A. Stability of mycotoxins during food processing. *Int. J. Food Microbiol.* **2007**, *119*, 140–146. [CrossRef] [PubMed]

23. Berthiller, F.; Crews, C.; Dall'Asta, C.; De Saeger, S.; Haesaert, G.; Karlovsky, P.; Oswald, I.P.; Seefelder, W.; Speijers, G.; Stroka, J. Masked mycotoxins: A review. *Mol. Nutr. Food Res.* **2013**, *57*, 165–186. [CrossRef] [PubMed]

24. Voss, K.A.; Snook, M.E. Stability of the mycotoxin deoxynivalenol (DON) during the production of flour-based foods and wheat flake cereal. *Food Addit. Contam.* **2010**, *27*, 1694–1700. [CrossRef] [PubMed]

25. Simsek, S.; Burgess, K.; Whitney, K.L.; Gu, Y.; Qian, S.Y. Analysis of deoxynalenol and deoxynivalenol-3-glucoside in wheat. *Food Control* **2012**, *26*, 287–292. [CrossRef]

26. Young, J.; Fulcher, R.; Hayhoe, J.; Scott, P.M.; Dexter, J. Effect of milling and baking on deoxynivalenol (vomitoxin) content of eastern Canadian wheats. *J. Agric. Food Chem.* **1984**, *32*, 659–664. [CrossRef]

27. Belitz, H.D.; Grosch, W.; Schieberle, P. *Lehrbuch der Lebensmittelchemie*, 6th ed.; Springer: Berlin/Heidelberg, Germany, 2008.

28. Zhang, H.J.; Wang, B.J. Fate of deoxynivalenol and deoxynivalenol-3-glucoside during wheat milling and Chinese steamed bread processing. *Food Control* **2014**, *44*, 86–91. [CrossRef]

29. Zhang, H.J.; Wang, B.J. Fates of deoxynivalenol and deoxynivalenol-3-glucoside during bread and noodle processing. *Food Control* **2015**, *50*, 754–757. [CrossRef]

30. Wu, L.; Wang, B.J. Evaluation on levels and conversion profiles of spiked DON, 3-ADON and 15-ADON during bread making process. *Food Chem.* **2015**, *185*, 509–516. [CrossRef]

31. Rani, K.U.; Prasada Rao, U.J.S.; Leelavathi, K.; Haridas Rao, P. Distribution of enzymes in wheat flour mill streams. *J. Cereal Sci.* **2001**, *34*, 233–242. [CrossRef]

32. Samar, M.M.; Neira, M.S.; Resnik, S.L.; Pacin, A. Effect of fermentation on naturally occurring deoxynivalenol (DON) in Argentinean bread processing technology. *Food Addit. Contam.* **2001**, *18*, 1004–1010. [CrossRef]

33. Carlier, L.; Baron, M.; Chamayou, A.; Couarraze, G. Greener pharmacy using solvent-free synthesis: Investigation of the mechanism in the case of dibenzophenazine. *Powder Technol.* **2013**, *240*, 41–47. [CrossRef]

34. Carlier, L.; Baron, M.; Chamayou, A.; Couarraze, G. Use of co-grinding as a solvent-free solid state method to synthesize dibenzophenazines. *Tetrahedron Lett.* **2011**, *52*, 4686–4689. [CrossRef]

Article

Selected Trichothecenes in Barley Malt and Beer from Poland and an Assessment of Dietary Risks Associated with their Consumption

Edyta Ksieniewicz-Woźniak [1], Marcin Bryła [1,*], Agnieszka Waśkiewicz [2], Tomoya Yoshinari [3] and Krystyna Szymczyk [1]

[1] Department of Food Analysis, Prof. Waclaw Dabrowski Institute of Agricultural and Food Biotechnology, Rakowiecka 36, 02-532 Warsaw, Poland; edyta.wozniak@ibprs.pl (E.K.-W.); krystyna.szymczyk@ibprs.pl (K.S.)

[2] Department of Chemistry, Poznan University of Life Sciences, Wojska Polskiego 75, 60-625 Poznan, Poland; agnieszka.waskiewicz@up.poznan.pl

[3] Division of Microbiology, National Institute of Health Sciences, 3-25-26 Tonomachi, Kawasaki-ku, Kawasaki-shi, Kanagawa 210-9501, Japan; t-yoshinari@nihs.go.jp

* Correspondence: marcin.bryla@ibprs.pl; Tel.: +48-22-606-3884

Received: 21 November 2019; Accepted: 6 December 2019; Published: 9 December 2019

Abstract: Eighty-seven samples of malt from several Polish malting plants and 157 beer samples from the beer available on the Polish market (in 2018) were tested for *Fusarium* mycotoxins (deoxynivalenol (DON), nivalenol (NIV)), and their modified forms ((deoxynivalenol-3-glucoside (DON-3G), nivalenol-3-glucoside (NIV-3G), 3-acetyldeoxynivalenol (3-AcDON)). DON and its metabolite, DON-3G, were found the most, among the samples analyzed; DON and DON-3G were present in 90% and 91% of malt samples, and in 97% and 99% of beer samples, respectively. NIV was found in 24% of malt samples and in 64% of beer samples, and NIV-3G was found in 48% of malt samples and 39% of beer samples. In the malt samples, the mean concentration of DON was 52.9 µg/kg (range: 5.3–347.6 µg/kg) and that of DON-3G was 74.1 µg/kg (range: 4.4–410.3 µg/kg). In the beer samples, the mean concentration of DON was 12.3 µg/L (range: 1.2–156.5 µg/L) and that of DON-3G was 7.1 µg/L (range: 0.6–58.4 µg/L). The concentrations of other tested mycotoxins in the samples of malt and beer were several times lower. The risk of exposure to the tested mycotoxins, following the consumption of beer in Poland, was assessed. The corresponding probable daily intakes (PDIs) remained a small fraction of the tolerable daily intake (TDI). However, in the improbable worst-case scenario, in which every beer bottle consumed would be contaminated with mycotoxins present at the highest level observed among the analyzed beer samples, the PDI would exceed the TDI for DON and its metabolite after the consumption of a single bottle (0.5 L) of beer.

Keywords: *Fusarium* toxins; modified mycotoxins; beer; malt; risk assessment

Key Contribution: High number of malt and beer samples were contaminated with mycotoxins. Strong beers (with higher alcohol content) contain higher levels of mycotoxins. Risk analysis showed a low level group probable daily intake of mycotoxin from beer. DON-3G present in beer has a significant share in group exposure to mycotoxins.

---

1. Introduction

Barley (*Hordeum vulgare* L.) has been grown for many years and is of great economic importance [1]. Approximately 57 million tonnes of barley was produced annually (in 2018) in the European Union, while global production has reached 147 million tonnes annually [2]. Most of the harvested grain is used as feed but the highest quality barley is selected for food production, including the production of malt. Malt is an ample source of the B-group vitamins, niacin, and minerals. It is increasingly used in the bakery and pastry industries to improve the quality of both the taste and health of their products [3]. However, beer production remains as the main application of malt [1,4]. Beer is an alcoholic beverage commonly consumed in numerous countries globally. Poland has the third largest quantity of beer production in Europe (approximately 93, 40.5, and 40.4 million hectoliters in Germany, UK, and Poland, respectively) and the fourth highest beer consumption per capita in Europe (approximately 138, 105, 101, and 97 liters in Czech Republic, Austria, Germany, and Poland, respectively [5]).

To arrive at a high-quality malt, one needs to start with a healthy grain with sufficiently high energy for germination and sufficient protein content. However, unfavorable climatic conditions during the plant vegetation season may negatively impact the quality of the grain and consequently, the decrease quality of the malt produced from that grain [6]. The most important climatic conditions are rainfall and temperature, which are two factors that mostly determine the degree to which the plants may become infected with pathogen fungi. *Fusarium* is one of the major fungal species infecting cereal grains, including barley. *Fusarium* head blight (FHB) disease caused by these fungi is a problem in various regions of the world. The fungal infection decreases crop yield, but even greater damage may result from the production of mycotoxins, which are secondary metabolites of the fungi that are toxic to humans and animals [7].

Fusarium spp. most often responsible for FHB in Poland include *F. graminearum*, *F. avenaceum*, and *F. culmorum*; however, other species are also seen in various regions of the world [8–10]. The mycotoxins produced by *Fusarium* in cereal grains include the trichothecenes, deoxynivalenol (DON), and nivalenol (NIV), and their modified forms. These toxins are also phytotoxic [11,12]. *F. culmorum* and *F. graminearum* are among the varieties that most aggressively infect plant ears [13,14]. Many of these fungi are capable of synthesizing 3- (3-AcDON) or 15-acetyl deoxynivalenol (15-AcDON), which are modified forms of DON [15]. Studies of the phytotoxic effects of DON have shown that the ability to covert DON into deoxynivalenol-3-glucoside (DON-3G) is the plant's primary defense mechanism against the toxin. Similar metabolic detoxication mechanisms help to build resistance to toxins in numerous cereal grain plants [16]. In barley, this mechanism is thought to be controlled by the QTL (quantitative trait loci)-specific region. Future studies involving deeper genetic analyses may help to develop tools to select fungal toxin-resistant plants using specific markers (marker-assisted selection; [17]). The phytotoxic effects of DON-3G are very weak compared to DON [18] and thus, it may be expected that a similar relationship holds for nivalenol 3-glucoside (NIV-3G) and NIV.

The consumption of DON- and/or NIV-contaminated food/feed may lead to disorders of the gastrointestinal tract, reproductive organs, and/or the immune system in both humans and animals. The toxicological characteristics of these toxins have been extensively described [19]. The lower levels of toxicity of DON-3G compared with DON have been confirmed in both humans and animals. In some in vitro studies and in some research on animals, it has been shown that DON-3G is not transported through the intestinal epithelium, but rather, is hydrolyzed by bacteria within the lower part of the alimentary tract [20]. Similar data are not available for NIV-3G, but it is commonly thought that the adverse effects of NIV-3G are weaker than those of NIV, as they are for DON-3G and DON.

Currently, the only European Commission regulation concerning mycotoxins in foodstuffs requires that the DON concentration in unprocessed cereal grains must not exceed 1250 µg/kg [21]. Taking into consideration the scientific evidence regarding the rapid absorption and excretion of DON, the in vivo deacetylation of 3- and 15-AcDON, and the hydrolysis of DON-3G in the lower parts of the alimentary tract; a European Food and Safety Authority (EFSA) expert panel recognized in 2017 that the toxic effects of DON-derivatives in humans may be comparable to the toxic effects of DON. Therefore, the tolerable daily intake (TDI) and reference dose (RfD) values have been recalculated as the sum of the three latter substances. Based on epidemiological data, a TDI threshold of 1 µg/kg body weight/day and an RfD dose of 8 µg/kg body weight/day have been accepted [19].

Reports on mycotoxins and their metabolites in Polish malts used in the brewing industry are very limited. The aims of this work included: (i) to assess the contamination of malts, sampled from several Polish malting plants, with selected *Fusarium* mycotoxins including their modified forms; (ii) to assess the mycotoxin contamination of beer available in 2019 on the Polish market; and (iii) to assess the risk of exposure to these mycotoxins following the consumption of beer in Poland.

2. Results and Discussion

2.1. Malt

Mycotoxins were found in the majority of the malt samples analyzed (Table 1). DON and DON-3G were found most often (in 90% and 91% of the malt samples, respectively) and at the highest levels (average of 52.9 and 74.1 µg/kg for DON and DON-3G, respectively). The percentage of samples positive for 3-AcDON was clearly lower 59% and NIV and NIV-3G were detected in the least number of samples (24% and 48%, respectively). DON-3G/DON molar ratios varied from 22% to 186% among DON-positive samples, while NIV-3G/NIV molar ratios varied from 32% to 126% among NIV-positive samples. Individual results regarding the content of individual mycotoxins in malt samples are presented in Table S1.

Table 1. Concentration of mycotoxins in 87 barley malt samples.

Assumed Values	Concentration (µg/kg)					Molar Ratios	
	DON	DON-3G	3-AcDON	NIV	NIV-3G	DON-3G /DON	NIV-3G /NIV
Positive samples (%)	78 (90%)	79 (91%)	51 (59%)	21 (24%)	42 (48%)	78 (90%)	21 (24%)
Average	52.9	74.1	7.7	22.1	13.9	89%	65%
Median	24.2	33.1	4.9	17.5	10.0	88%	66%
Min–Max	5.3–347.6	4.4–410.3	2.2–40.2	8.3–118.6	5.0–57.4	22%–186%	32%–126%

DON, deoxynivalenol; DON-3G, deoxynivalenol-3-glucoside; 3-AcDON, 3-acetyldeoxynivalenol; NIV, nivalenol; NIV-3G, nivalenol-3-glucoside.

In grains, DON-3G is known to be a product of the plant defense reaction to the presence of the phytotoxin, DON [22–24]. DON-3G is easily soluble and plants can easily transport it from the cytoplasm to vacuoles or the intercellular space [16]. The DON-3G/DON ratio in the grain itself does not usually exceed 30% [25,26]. However, in malt samples we observed an average DON-3G/DON ratio of 89%, with a range of 22%–186%. Relatively high values (average 65%, range 32%–126%) were also noted for the NIV-3G/NIV ratio. Some researchers have suggested that changes occur during the malting process that activate secondary detoxicating enzymes, which then catalyze the conversion of the toxins to their glycoside derivatives [27–29]. Maul et al. [29] have shown that sprouting seeds of barley, millet, oat, rye, and spelt are capable of converting DON into DON-3G by means of UDP-glucosyltransferases. In barley, approximately 50% of DON was found to be converted, mainly into DON-3G, with a similar conversion rate observed in wheat. Moreover, Lancova et al. [28] reported that, during barley grain germination, the concentration of DON may decrease by 90%, while the concentration of DON-3G may markedly increase, to a level as high or several times higher than DON. Spanic et al. [30] presented data on mycotoxin levels in wheat varieties varying in Fusarium head blight resistance; the average content of DON-3G increased from 59.9 µg/kg in grain to 163.9 µg/kg in malt.

There are very few reports in the literature on the co-occurrence of DON/DON-3G and NIV/NIV-3G in brewing malts, even though such data are essential for regulating food safety. In the present study, we detected these substances in both malt and beer samples. However, the DON concentration did not exceed 750 µg/kg, the maximum permissible level in malt specified in EC Regulation 1881/2006, in any of the tested malt samples [21]. Practically, malt plants in Poland do not purchase grain contaminated with DON at levels above 1 mg/kg, while the maximum permissible level in grain is 1.25 mg/kg, as per EC Regulation 1881/2006 [21]. Mitteleuropäische Brautechnische Analyskomommision [31] recommends the inspection of each batch of grain offered to a malting plant for the presence of *F. graminearum* and

F. culmorum. If mycelia are visible, they recommend the analysis of the grain for mycotoxins. There are some indications in the literature [32–35] that high amounts of additional mycotoxins may be synthesized in fungi-contaminated grain during the malting process, thus significantly impacting food safety.

2.2. Beer

The majority of beers marketed in Poland are light beers based on pilsner malts. However, dark ale or lager beers produced from Munich malts, usually obtained from lower quality grains [36], caramel malts or roasted pale ale malts are also popular. The two latter malts are enzymatically inactive; they are introduced in small amounts [37], to darken the beer and enhance its flavor. Wheat beers are also becoming increasingly common on the market. They are produced from barley malt, with the addition of at least 50% wheat or wheat malt. The flavor of these beers is unique, differing from the flavor of classical barley-only beers [38]. We divided our beer samples into three common categories for analysis: light, dark, and wheat beers. The percentage of mycotoxin-positive beer samples in all these groups was high (Table 2). Individual results regarding the content of mycotoxins in beer samples are presented in Table S1.

Table 2. Concentration of mycotoxins in light, dark, and wheat beers.

Type of Beer		Concentration (µg/L)					Molar Ratios	
		DON	DON-3G	3-AcDON	NIV	NIV-3G	DON-3G /DON	NIV-3G /NIV
Light beers (*n* = 105)	No. of positive samples (%)	101 (96%)	103 (98%)	72 (69%)	70 (67%)	45 (43%)	100 (95%)	42 (40%)
	Average	13.0	7.3	1.0	1.5	1.1	46%	42%
	Median	8.0	4.8	0.7	1.4	0.8	33%	30%
	Min–Max	1.2–156.5	0.6–36.8	0.3–8.3	0.6–3.6	0.5–4.5	10–149%	12–137%
Dark beers (*n* = 28)	No. of positive samples (%)	28 (100%)	28 (100%)	7 (25%)	15 (54%)	7 (25%)	28 (100%)	6 (21%)
	Average	11.7	7.8	1.2	1.0	0.7	40%	41%
	Median	8.8	4.8	0.8	0.8	0.6	39%	36%
	Min–Max	2.7–54.4	1.3–58.4	0.3–3.9	0.6–2.5	0.5–0.8	18–71%	30–74%
Wheat beers (*n* = 24)	No. of positive samples (%)	24 (100%)	24 (100%)	13 (58%)	15 (63%)	10 (42%)	24 (100%)	9 (38%)
	Average	9.6	5.0	0.9	1.1	0.9	34%	50%
	Median	9.9	3.8	0.9	1.0	0.9	34%	52%
	Min–Max	2.2–24.6	0.6–13.2	0.9–1.9	0.6–2.0	0.5–1.6	14–59%	23–79%
Total (*n* = 157)	No. of positive samples (%)	153 (97%)	155 (99%)	92 (59%)	100 (64%)	62 (39%)	152 (97%)	57 (36%)
	Average	12.3	7.1	1.0	1.3	1.1	43%	43%
	Median	8.6	4.8	0.8	1.2	0.8	38%	37%
	Min–Max	1.2–156.5	0.6–58.4	0.3–8.3	0.6–3.6	0.5–4.5	10–149%	12–137%

DON, deoxynivalenol; DON-3G, deoxynivalenol-3-glucoside; 3-AcDON, 3-acetyldeoxynivalenol; NIV, nivalenol; NIV-3G, nivalenol-3-glucoside.

As was the case for malt samples, DON and DON-3G were the most frequently found toxins in beer samples, being present in 96% and 98% of light beer samples, respectively, and in all the samples of dark and wheat beers. Other mycotoxins, namely, 3-AcDON, NIV, and NIV-3G were found at lower levels in 69%, 25%, and 58%; 67%, 54%, and 63%; and 43%, 25%, and 42% of the light, dark, and wheat beer samples, respectively. The maximum DON (156.5 µg/L) and DON-3G (58.4 µg/L) concentrations were found in a light and a dark beer sample, respectively. The average levels of the three remaining tested mycotoxins ranged from 0.7 to 1.5 µg/L, i.e., they were approximately 6–20 times lower than the DON levels. The average DON-3G/DON and NIV-3G/NIV molar ratios ranged from 34% to 46% and 41% to 50%, respectively. Neither the mycotoxin concentrations nor their molar ratios were dependent on the beer category.

The alcohol content of beer depends on the extent to which the yeast ferments the sugars, which largely depends on the amount of grain and malt in the fermentation batch. Stronger beer requires more grain, which results in a higher risk of mycotoxin contamination [27,39,40]. Grain extracts used for beer production contain mainly sugars but may also contain dextrins, nitrogenous compounds (proteins), mineral salts, and other compounds, depending on the recipe used by the beer manufacturer [41]. Therefore, a comparison of the level of mycotoxin contamination in beers with different extract contents must be treated only as an approximation. Therefore, we re-organized the beer samples into three different categories: mild beers (0.5–5.0% alcohol, 3.5–12.5% extract), regular beers (5.1–6.0% alcohol, 6.8–16.0% extract), and strong beers (6.1–10.0% alcohol, 8.4%–21.0% extract; Table 3).

Table 3. Concentrations of mycotoxins in mild, regular, and strong beers.

Type of Beer		Concentration (µg/L)					Molar Ratios	
		DON	DON-3G	3-AcDON	NIV	NIV-3G	DON-3G /DON	NIV-3G /NIV
Mild beers *n* = 48)	No. of positive samples (%)	45 (94%)	47 (98%)	26 (54%)	21 (44%)	18 (38%)	45 (94%)	15 (31%)
	Average	7.1	5.6	0.8	1.2	1.3	50%	58%
	Median	4.3	3.0	0.7	1.0	1.0	45%	43%
	Min–Max	1.4–24.6	0.6–30.9	0.3–2.7	0.6–2.3	0.5–4.5	18–149%	26–137%
Regular beers (*n* = 61)	No. of positive samples (%)	61 (100%)	61 (100%)	40 (67%)	45 (75%)	23 (38%)	59 (97%)	23 (38%)
	Average	12.1	7.0	0.9	1.5	1.1	42%	41%
	Median	9.5	5.2	0.8	1.3	0.9	37%	37%
	Min–Max	1.2–54.2	0.6–31.5	0.4–2.6	0.6–3.6	0.5–2.8	15–118%	16–90%
Strong beers (*n* = 48)	No. of positive samples (%)	48 (100%)	48 (100%)	27 (56%)	34 (71%)	21 (44%)	48 (100%)	19 (40%)
	Average	17.3	8.6	1.3	1.3	0.7	38%	34%
	Median	8.5	5.2	0.9	1.1	0.7	37%	32%
	Min–Max	2.0–156.5	0.6–58.4	0.3–8.3	0.6–3.3	0.5–1.8	10–104%	12–52%

DON, deoxynivalenol; DON-3G, deoxynivalenol-3-glucoside; 3-AcDON, 3-acetyldeoxynivalenol; NIV, nivalenol; NIV-3G, nivalenol-3-glucoside.

The number of positive samples and the concentration of the majority of the tested mycotoxins positively correlated with alcohol content in most cases. DON and DON-3G were the predominant toxins in 94% and 98% of mild beer samples, respectively, and in all samples of regular and strong beer, with average DON concentrations of 7.1, 12.1, and 17.3 µg/L and average DON-3G concentrations of 5.6, 7.0, and 8.6 µg/L for mild, regular, and strong beers, respectively. Less clear, but similar trends were noted for the other tested mycotoxins.

Mycotoxin contamination of beer has been studied by numerous groups (Table 4). However, data on the co-occurrence of DON, DON-3G, 3-AcDON, NIV, and NIV-3G in beer are scarce. The scope of most reported studies has been restricted to DON, DON-3G, and 3-AcDON, with a few studies also including NIV. Typically, the reported concentrations of the predominant DON have not exceeded 100 µg/L [27,42–45]. The findings from the present study mostly agree with those from previous studies (because the fraction of positive samples may depend on the LOD and LOQ of the method used). Higher concentrations of DON have been found mainly in beers originating from non-European countries, including craft beers from Brazil (127–501 µg/L; [46]), traditional African beers from Cameroon (140–730 µg/L; [47]), and Busaa-type beers from Kenya (200–360 µg/kg [48]). However, relatively high DON concentrations (104–182 µg/L) have also been found in strong (>8% alcohol) Norwegian Imperial Stout beer [49]. In this study, we found a high DON concentration (156.5 µg/L) only in one strong (>8% alcohol) sample of a light beer.

Table 4. Selected literature data on mycotoxins in beer.

Beer	No. of Samples	Toxin	LOD (µg/L)	LOQ (µg/L)	Concentration(µg/L)		Reference
					Average	Max	
Wheat beer	46	DON	1	4.5	18.4	49.6	[43]
		DON-3G	0.9	3.5	11.5	28.4	
		3-AcDON	2.2	8.2	<LOD	<LOD	
Pale beer	217	DON	2.2	5.4	12	89.3	
		DON-3G	0.4	3.5	9.3	81.3	
		3-AcDON	2.4	6.8	<LOD	<LOD	
Dark beer	47	DON	2.9	11	22.4	45	
		DON-3G	1.4	4.1	10.7	26.2	
		3-AcDON	4.3	11	<LOD	<LOD	
Bock beer	20	DON	1.2	4.1	13.8	27.1	
		DON-3G	0.5	1.5	14.8	33.3	
		3-AcDON	3.6	9.2	<LOD	<LOD	
Non-alcoholic beer	19	DON	1.2	3	14.8	33.3	
		DON-3G	0.4	1.4	3	6.6	
		3-AcDON	2.6	6	<LOD	<LOD	
Shandy beer	25	DON	1.5	3.9	6.9	12.7	
		DON-3G	0.4	1.3	3.8	7.9	
		3-AcDON	2.7	10	<LOD	<LOD	
Wheat beer	10	DON	1	4.5	14	27	[44]
		DON-3G	0.9	3.5	8.6	15	
		3-AcDON	2.2	8.2	<LOD	<LOD	
Pale beer	10	DON	2.2	5.4	13	30	
		DON-3G	0.4	3.5	8.3	19	
		3-AcDON	2.4	6.8	<LOD	<LOD	
Dark beer	10	DON	2.9	11	11	11	
		DON-3G	1.4	4.1	9.6	16	
		3-AcDON	4.3	11	<LOD	<LOD	
Bock beer	10	DON	1.2	4.1	13	22	
		DON-3G	0.5	1.5	16	32	
		3-AcDON	3.6	9.2	<LOD	<LOD	
Non-alcoholic beer	10	DON	1.2	3	3.7	3.7	[44]
		DON-3G	0.4	1.4	2.3	3.1	
		3-AcDON	2.6	6	<LOD	<LOD	
Shandy beer	10	DON	1.5	3.9	6.4	6.4	
		DON-3G	0.4	1.3	3.5	5.5	
		3-AcDON	2.7	10	<LOD	<LOD	
Light beers	158	DON	1	2.5	1.6–9.2 (depending on alcohol content)	3.7–35.9	[27]
		DON-3G	1	2.5	1.7–5.8	1.2–37	
		AcDONs	2	5	1.7–5.8	1.0–25	
		NIV	2.5	10	<LOD	<LOD	
Dark beers	18	DON	1	2.5	1.3–11.2	1.0–16.0	
		DON-3G	1	2.5	<LOQ–7.8	<LOQ–26.0	
		AcDONs	2	5	<LOQ–13.7	<LOQ–24.0	
		NIV	2.5	10	<LOD	<LOD	

Table 4. *Cont.*

Beer	No. of Samples	Toxin	LOD (µg/L)	LOQ (µg/L)	Concentration(µg/L)		Reference
					Average	Max	
African traditional beer	10	DON	n.r	10	81.8	140	
		DON-3G		2.5	<LOD	<LOD	
		AcDONs		10	<LOQ	<LOQ	
		NIV		5	8.7	9	
Bock beer	2	DON	n.r.	10	52	64	
		DON-3G		2.5	60	97	
		AcDONs		10	<LOD	<LOD	
		NIV		5	<LOD	<LOD	[49]
Dark lager	2	DON	n.r.	10	32.5	41	
		DON-3G		2.5	52	68	
		AcDONs		10	<LOD	<LOD	
		NIV		5	<LOD	<LOD	
Double India Pale Ale	1	DON	n.r.	10	67	67	
		DON-3G		2.5	48	48	
		AcDONs		10	<LOD	<LOD	
		NIV		5	<LOD	<LOD	
Eisbock	1	DON	n.r.	10	32	32	
		DON-3G		2.5	32	32	
		AcDONs		10	<LOD	<LOD	
		NIV		5	<LOD	<LOD	
Fruit/Vegetable/ Spice	1	DON	n.r.	10	<LOQ	<LOQ	
		DON-3G		2.5	LOD	LOD	
		AcDONs		10	<LOD	<LOD	
		NIV		5	<LOD	<LOD	
Imperial Stout	18	DON	n.r.	10	95.1	412	
		DON-3G		2.5	96.7	619	
		AcDONs		10	<LOD	<LOD	
		NIV		5	<LOD	<LOD	
India Pale Ale	3	DON	n.r.	10	40	64	
		DON-3G		2.5	14	18	
		AcDONs		10	<LOD	<LOD	[49]
		NIV		5	<LOD	<LOD	
Non/Low Alcohol	1	DON	n.r.	10	<LOQ	<LOQ	
		DON-3G		2.5	<LOD	<LOD	
		AcDONs		10	<LOD	<LOD	
		NIV		5	<LOD	<LOD	
Pale ale	5	DON	n.r.	10	20.3	40	
		DON-3G		2.5	29.5	82	
		AcDONs		10	<LOQ	<LOQ	
		NIV		5	<LOD	<LOD	
Pale Lager	6	DON	n.r.	10	12.5	13	
		DON-3G		2.5	22	53	
		AcDONs		10	<LOQ	<LOQ	
		NIV		5	<LOD	<LOD	
Smoked	1	DON	n.r.	10	23	23	
		DON-3G		2.5	14	14	
		AcDONs		10	<LOQ	<LOD	
		NIV		5	<LOD	<LOD	

Table 4. *Cont.*

Beer	No. of Samples	Toxin	LOD (µg/L)	LOQ (µg/L)	Concentration(µg/L)		Reference
					Average	Max	
Sour Ale	4	DON	n.r.	10	17	29	
		DON-3G		2.5	16.7	22	
		AcDONs		10	<LOD	<LOD	
		NIV		5	<LOD	<LOD	
Stout	4	DON	n.r.	10	28	30	
		DON-3G		2.5	41.3	52	
		AcDONs		10	<LOD	<LOD	
		NIV		5	<LOD	<LOD	
Strong Dark Pale	3	DON	n.r.	10	17.5	25	
		DON-3G		2.5	26.5	35	
		AcDONs		10	<LOQ	<LOD	
		NIV		5	<LOD	<LOD	[49]
Strong Pale Ale	9	DON	n.r.	10	17.5	25	
		DON-3G		2.5	26.5	35	
		AcDONs		10	<LOQ	<LOD	
		NIV		5	<LOD	<LOD	
Strong Pale Lager	1	DON	n.r.	10	12	12	
		DON-3G		2.5	17	17	
		AcDONs		10	<LOQ	<LOD	
		NIV		5	<LOD	<LOD	
Wheat beer	5	DON	n.r.	10	10	32	
		DON-3G		2.5	4	41	
		AcDONs		10	<LOQ	<LOD	
		NIV		5	<LOD	<LOD	
Mild beer	28	DON	1.3	4.1	10.5	65	
		DON-3G	1.9	6.2	7.6	25	
		NIV	0.6	2.1	2.7	4.8	
Regular beer	34	DON	1.3	4.1	6.6	19.7	
		DON-3G	1.9	6.2	8.8	35.8	[42]
		NIV	0.6	2.1	1.5	7.4	
Strong beer	38	DON	1.3	4.1	10	73.6	
		DON-3G	1.9	6.2	10.3	35.2	
		NIV	0.6	2.1	2.8	7.6	

n.r. = not reported.

Some of the beer samples tested had a higher concentration of DON-3G than DON. Similar DON-3G/DON molar ratios have been reported in the literature, with averages of 0.56 (range 0.11–1.25 [43]) and 0.79 (range 0.1–2.6 [49] and 0.7–1.0 [26]. As can be seen, the DON-3G/DON molar ratios in beer are similar to those in malt.

2.3. Dietary Exposure Assessment

The following group TDI values were used in the assessment of risk of exposure to mycotoxins following beer consumption: 1 µg/kg body weight/day of the sum of DON, DON-3G, 3-AcDON, and 15-AcDON [20] and 1.2 µg/kg body weight/day of the sum of NIV and NIV-3G [50]. The average beer consumption in Poland is 97 L per capita annually, i.e., 0.27 L per capita per day [5]. In three considered scenarios, it was assumed that consumed beer contained mycotoxins at a level equal to: (i) the median, (ii) the third quartile, or (iii) the maximum concentration found in our samples (the worst-case scenario). It was assumed that the average adult in Poland weighs 70 kg. The results of the calculations are shown in Table 5. PDI values remained a small fraction of TDI values in the first and second scenarios (5.1% and 7.9%, respectively, for DON and its derivatives and 0.32% and 0.61%, respectively, for NIV and its derivatives). In the improbable third scenario (worst case), the PDI would reach 65.2% of the TDI for DON and its derivatives and 2.41% of the TDI for NIV and its derivatives.

Table 5. Group probable daily intake and its share of the total daily intake calculated in three scenarios, in which different concentrations of mycotoxins were assumed in the consumed beer.

Assumed Values	DON			DON+DON3G+3AcDON			NIV+NIV3G		
	Concentration (µg/L)	* PDI (ng/kg b.w./day)	%TDI	Concentration (µg/L)	PDI (ng/kg b.w./day)	%TDI	Concentration (µg/L)	PDI (ng/kg b.w./day)	%TDI
Median **	8.3	31.5	3.2	14.2	50.7	5.1	1.1	3.8	0.32
Quartile 3 **	13.3	50.4	5.0	22.2	79.4	7.9	2.1	7.3	0.61
Maximum	156.5	594.1	59.4	182.5	651.8	65.2	8.1	28.9	2.41

DON, deoxynivalenol; DON-3G, deoxynivalenol-3-glucoside; 3-AcDON, 3-acetyldeoxynivalenol; NIV, nivalenol; NIV-3G, nivalenol-3-glucoside; PDI, probable daily intake; TDI, total daily intake; * PDI = $\frac{C \cdot C_d}{b.w.}$, where C is concentration of the mycotoxin in the contaminated beer, C_d is the average daily consumption of beer in Poland, and b.w. is mean body weight. ** If the measurement for any analyte was below the LOQ, the median and 3rd quartile were calculated assuming that the analyte was present at the level of LOQ/2.

The average consumption of 0.27 L of beer per day assumed in the above dietary exposure assessment does not reflect the real situation, since beer consumers rarely drink less than one bottle (0.5 L) per day. The PDI for persons drinking 0.5 L of beer daily would be approximately twice the values calculated above, in which case the TDI of DON and its derivatives would exceed the worst-case scenario by approximately 30%. Each additional beer bottle consumed per day would double the above calculated PDI values. It is also worth noting that the analytical method developed here was not efficient at detecting 15-AcDON. However, since 3-AcDON was detected at very low levels, one can expect that the contribution of 15-AcDON to the PDI is insignificant.

Of course, beer is not the main source of DON and its derivatives (the most important trichothecenes from a food safety point of view) in the human diet. Greater levels of exposure come from the consumption of bakery products, corn flakes, pasta, and other grain-based foodstuffs that are consumed daily, not only by beer consumers. Considering the exposition, bakery products and pastas are in Europe more and more often indicated as a possible quite serious threat to human health [50]. Studies of markers in urine have shown that chronic exposure to DON and its derivatives is greater than the accepted TDI [51–53]. Therefore, the consumption of beer may increase the risk of excessive mycotoxin exposure.

Data on the risks associated with the consumption of mycotoxin-contaminated beers exist only with respect to officially regulated toxins. It is a common observation that DON is the greatest risk factor, but beer is not generally considered an important source of dietary mycotoxin exposure. Even if the maximum detected DON concentrations are taken into account, the PDI values remain a small percentage of the TDI values, regardless of the country of origin of the beer. For example, the PDI is 14.0–20.8% of the TDI in Poland [54]; 18% of the TDI in Brazil [46]; 0.15–6.14% of the TDI in Spain, where the average consumption is just half of that in Poland [55]; 0% of the TDI in Cyprus and 10% of the TDI in Ireland [56].

The consumption of mycotoxin-contaminated beer results in negligible risk of exposure to NIV and NIV-3G. EFSA has reported that even the consumption of bakery products and pasta is safe in terms of exposure to these toxins [57]. In view of the low concentrations of NIV and NIV-3G, the PDI values are far below the TDI values, even for foodstuffs that are consumed in relatively large quantities, such as bakery products and pasta.

3. Conclusions

The data presented here on the co-occurrence of DON, NIV, and their metabolized (masked) forms in brewing malts and beers available on the Polish market are among the first reported in the literature. Mycotoxins were found in the majority of the barley malt and beer samples tested. DON and its metabolite, DON-3G, were found most frequently (in more than 90% of samples), although at safely low levels. NIV and its metabolite, NIV-3G, were found at lower levels in malt and beer samples. Because of the low mycotoxin levels, none of the tested beers were regarded as unsafe from a toxicological point of view. However, in the worst-case scenario, the PDI would exceed the TDI for DON and its metabolites after drinking just one bottle (0.5 L) of beer.

4. Materials and Methods

4.1. Reagents and Standards

Certified reference standards of DON, 3-AcDON, and NIV (100 µg/mL in acetonitrile), and DON-3G (50 µg/mL in acetonitrile:water, 50:50, *v/v*), were purchased from Romer Labs (Tulln, Austria). NIV-3G (110 µg/mL) was isolated from wheat, according to the procedure described by Yoshinari et al. [58]. Acetonitrile, methanol, and LC/MS-grade water were purchased from Witko (Łódź, Poland). Ammonium formate and formic acid (LC-MS grade) were obtained from Fisher Scientific (Millersburg, PA, USA). DON-NIV wide-bore (WB) immunoaffinity columns and PBS buffer solutions were purchased from Vicam (Watertown, NY, USA).

4.2. Research Material

One hundred and fifty-seven beer samples and 87 barley malt samples were analyzed. Various brands of light, dark, and wheat beers (mild, regular, and strong) were purchased in 2019 from local supermarkets in Poland. Malt was sampled from various malt plants located throughout the country, in line with the guidelines specified within EC Regulation 519/2014 (February 23, 2006) [59], which describes sampling and analysis methods for the official control of mycotoxin levels in foodstuffs. All the acquired samples belonged to the most common Pilsner malts, which are used to produce pale straw-colored ale and lager beers [36]. Malt samples, each with a mass of approximately 1 kg, were ground in a Knife Mill Grindomix GM 200 grinder (Retsch GmbH, Haan, Germany).

4.3. Sample Preparation

Malt and beer samples were prepared for analysis using a method previously described by our research team [42,60]. After extraction and homogenization (for malt extraction in Unidrive 1000 homogenizer, CAT Scientific Inc., Paso Robles, CA, USA), each sample was passed through a DON-NIV WB immunoaffinity column at a speed of 1–2 drops/s. The column was rinsed with 10 mL of PBS and 10 mL of de-ionized water. Analytes were washed out of the column, first with 0.5 mL of methanol and then with 1.5 mL of acetonitrile and were collected into a reaction vial. The solvent was evaporated in a stream of nitrogen. The residues were re-dissolved in 300 µL of 30% methanol and analyzed by liquid chromatography-mass spectrometry (LC-MS). Samples were analyzed at three replications.

4.4. LC-MS Analysis

An H-class liquid chromatograph coupled to a mass spectrometer with a time-of-flight analyzer (UPLC-TOF-HRMS; Waters, Milford, MA, USA) was used to analyze mycotoxins. Analytes were separated on a 2.1 × 100 mm, 1.6 µm UPLC C18 Cortecs chromatographic column (Waters) with an appropriate pre-column, operated with a gradient regime. Phase A was 90:10 *v/v* methanol:water, phase B was 10:90 *v/v* methanol:water. Both phases contained 0.2% formic acid and 10 mM ammonium formate. The flow rate was 0.3 mL/min, with the following flow gradient: 0–2 min, 100% B; 3–6 min, 50% B; 22–23 min, 100% A; and 25–28 min, 100% B. Five microliters of each sample was injected onto the column. The mass spectrometer was operated in the positive/negative electrospray ionization mode, with an ion source temperature of 150 °C and a desolvation temperature of 300/350 °C for positive/negative ionization, respectively. The nebulizing gas (N_2) flow rate was 750 L/min and the cone gas flow rate was 40 L/min. The capillary bias was 3200 V. Ion optics was operated in V mode and the instrument was calibrated using a leucine-enkephalin solution.

4.5. Method Validation

Linearity ranges, limits of detection (LOD, the concentration at which the signal:noise ratio was 3), limits of quantification (LOQ, the concentration at which the signal:noise ratio was 10), recovery rates (R), and repeatability/precision (expressed as the relative standard deviation [RSD]), were determined using calibration curves that were constructed using separate blank samples for each mycotoxin of interest in the beer and malt matrices. The blanks were prepared in the same way as the analytes, except that the respective amount of standard mixture was added just prior to finally dissolving it in 30% methanol, after which the solvent was removed in a dry nitrogen stream. Each calibration curve consisted of eight points. The concentrations covered for the malt samples (in µg/kg) were: 5.0–1028 for DON; 4.0–516 for DON-3G; 2.0–1028 for 3-AcDON; 8.0–1050 for NIV; and 5.0–565 for NIV-3G. The concentrations covered for the beer samples (in µg/L) were: 3–68.6 for DON; 2.1–34.4 for DON-3G; 0.9–68.6 for 3-AcDON; 2.1–70.1 for NIV; and 1.6–37.7 for NIV-3G. The results of the analytical method validation experiment are shown in Tables 6 and 7.

Table 6. Limits of detection, limits of quantification, and determination coefficients for individual analytes determined in malt and beer samples.

Analyte	Ion Mass (*m/z*)	Retention Time (min)	Malt			Beer		
			LOD (µg/kg)	LOQ (µg/kg)	R^2	LOD (µg/L)	LOQ (µg/L)	R^2
DON	341.2 (M+FA−H)⁻	4.08	5	17	0.9891	0.6	2.1	0.9977
DON-3G	503.2 (M+FA−H)⁻	4.22	4	13	0.9910	0.5	1.6	0.9919
3-AcDON	339.2 (M+H)⁺	4.98	2	7	0.9974	0.3	0.9	0.9899
NIV	357.2 (M+FA−H)⁻	2.38	8	24	0.9909	1.0	3.0	0.9889
NIV-3G	519.2 (M+FA−H)⁻	2.45	5	17	0.9905	0.6	2.1	0.9989

DON, deoxynivalenol; DON-3G, deoxynivalenol-3-glucoside; 3-AcDON, 3-acetyldeoxynivalenol; NIV, nivalenol; NIV-3G, nivalenol-3-glucoside; LOD, limit of detection; LOQ, limit of quantification; R^2, determination coefficient.

Table 7. Recovery rates and relative standard deviations for individual analytes determined in malt and in beer samples spiked at different fortification levels.

Analyte	Malt (*n* = 4)			Beer (*n* = 4)		
	Fortification Level (µg/kg)	R (%)	RSD (%)	Fortification Level (µg/L)	R (%)	RSD (%)
DON	42.9	90.7	12.9			
	128.6	94.3	8.2	17.1	75.0	8.8
	514.3	101.3	11.9	34.3	106.0	2.8
	1028.5	97.0	15.0	68.6	85.0	9.5
DON-3G	21.5	87.4	6.1			
	64.5	73.5	9.8	8.6	87.0	6.7
	258.1	89.9	11.1	17.2	93.0	2.5
	516.2	79.1	15.4	34.4	89.0	6.0
3-AcDON	42.9	105.1	18.4			
	128.6	105.4	4.7	17.1	93.0	6.7
	514.3	103.8	8.8	34.3	97.0	2.7
	1028.5	102.1	22.1	68.6	87.0	6.7
NIV	43.8	89.9	11.4			
	131.4	85.8	8.6	17.5	80.0	6.2
	525.6	85.1	9.3	35.0	100.0	6.5
	1051.2	83.7	13.0	70.1	91.0	9.4
NIV-3G	23.6	105.0	13.7			
	70.7	85.0	7.8	9.4	93.0	6.7
	282.8	87.7	9.6	18.9	101.0	6.6
	565.6	86.4	13.1	37.7	96.0	8.0

DON, deoxynivalenol; DON-3G, deoxynivalenol-3-glucoside; 3-AcDON, 3-acetyldeoxynivalenol; NIV, nivalenol; NIV-3G, nivalenol-3-glucoside; R, recovery rate; RSD, relative standard deviation.

Since all analytes of interest belonged to the trichothecenes group, we assessed the performance of the method for DON analysis using the following specifications listed in EC Regulation 519/2014 [59]: recovery rates 60%–110% or 70%–120%, depending on the fortification level and RSD ≤20%. These criteria were met in 34 out of 35 analyte/fortification level combinations. In one case, the RSD was above 20%.

This validated method was then used to analyze DON, DON-3G, 3-AcDON, NIV, and NIV-3G in the malt and beer samples.

Supplementary Materials: The following are available online at http://www.mdpi.com/2072-6651/11/12/715/s1, Table S1: Individual results of mycotoxin concentrations in the analyzed beer and malt samples.

Author Contributions: Conceptualization, E.K.-W. and M.B.; methodology, M.B. and A.W.; formal analysis, M.B. and E.K.-W.; performed isolation of the NIV-3G analytical standard, T.Y.; performed the manuscript preparation, M.B., E.K.-W. and T.Y.; supervised the research, A.W., K.S.

Funding: This research was financially supported from the Polish National Science Centre project 2016/21/D/NZ9/02597.

Conflicts of Interest: The authors declare no conflict of interest.

References

1. Zhou, M.X. Barley Production and Consumption. In *Genetics and Improvement of Barley Malt Quality*; Zhang, G., Li, C., Eds.; Springer: Berlin/Heidelberg, Germany, 2009; pp. 1–17.

2. Central Statistics Office (CSO). Available online: https://www.cso.ie/en/releasesandpublications/er/aypc/areayieldandproductionofcrops2018 (accessed on 20 September 2019).

3. Ha, K.S.; Jo, S.H.; Mannam, V.; Kwon, Y.I.; Apostolidis, E. Stimulation of phenolics, antioxidant and α-glucosidase inhibitory activities during barley (*Hordeum vulgare* L.) seed germination. *Plant Food Hum. Nutr.* **2016**, *71*, 211–217. [CrossRef]

4. Tricase, C.; Amicarelli, V.; Lamonaca, E.; Rana, R.L. Economic analysis of the barley market and related uses. In *Grasses as Food and Feed*; Tadele, Z., Ed.; IntechOpen Limited: London, UK, 2018; pp. 25–46.

5. The Brewers of Europe. *Beer Statistics*, 2018 ed.; The Brewers of Europe: Brussels, Belgium, 2018.

6. Eagles, H.A.; Bedggood, A.G.; Panozzo, J.F.; Martin, P.J. Cultivar and environmental effects on malting quality in barley. *Aust. J. Agric. Res.* **1995**, *46*, 831–844. [CrossRef]

7. Popovski, S.; Celar, F.A. The impact of environmental factors on the infection of cereals with *Fusarium* species and mycotoxin production-a review. *Acta Agric. Slov.* **2013**, *101*, 105–116. [CrossRef]

8. Perkowski, J.; Kiecana, I.; Kaczmarek, Z. Natural occurrence and distribution of *Fusarium* toxins in contaminated barley cultivars. *Eur. J. Plant. Pathol.* **2003**, *109*, 331–339. [CrossRef]

9. Nielsen, L.K.; Cook, D.J.; Edwards, S.G.; Ray, R.V. The prevalence and impact of Fusarium head blight pathogens and mycotoxins on malting barley quality in UK. *Int. J. Food Microbiol.* **2014**, *179*, 38–49. [CrossRef] [PubMed]

10. Schwarz, P.B.; Schwarz, J.G.; Zhou, A.; Prom, L.K.; Steffenson, B.J. Effect of *Fusarium graminearum* and *F. Poae* infection on barley and malt quality. *Monatsschrift Brauwiss.* **2001**, *54*, 55–63.

11. Abbas, H.K.; Yoshizawa, T.; Shier, W.T. Cytotoxicity and phytotoxicity of trichothecene mycotoxins produced by *Fusarium* spp. *Toxicon* **2013**, *74*, 68–75. [CrossRef] [PubMed]

12. International Agency for Research on Cancer. *Some Naturally Occurring Substances: Food Items and Constituents, Heterocyclic Aromatic Amines and Mycotoxins*; Monograph on the evaluation of carcinogenic risks to humans; International Agency for Research on Cancer: Lyon, France, 1993.

13. Pestka, J.J. Deoxynivalenol: Toxicity, mechanisms and animal health risks. *Anim. Feed Sci. Technol.* **2007**, *137*, 283–298. [CrossRef]

14. Zinedine, A.; Soriano, J.M.; Moltó, J.C.; Mañes, J. Review on the toxicity, occurrence, metabolism, detoxification, regulations and intake of zearalenone: An oestrogenic mycotoxin. *Food Chem. Toxicol.* **2007**, *45*, 1–18. [CrossRef]

15. Pasquali, M.; Beyer, M.; Logrieco, A.; Audenaert, K.; Balmas, V.; Basler, R.; Boutigny, A.L.; Chrpová, J.; Czembor, E.; Gagkaeva, T.; et al. A European database of *Fusarium graminearum* and *F. culmorum* trichothecene genotypes. *Front. Microbiol.* **2016**, *7*, 406. [CrossRef]

16. Shin, S.; Torres-Acosta, J.A.; Heinen, S.J.; McCormick, S.; Lemmens, M.; Kovalsky Paris, M.P.; Berthiller, F.; Adam, G.; Muehlbauer, G.J. Transgenic *Arabidopsis thaliana* expressing a barley UDP-glucosyltransferase exhibit resistance to the mycotoxin deoxynivalenol. *J. Exp. Bot.* **2012**, *63*, 4731–4740. [CrossRef] [PubMed]

17. Miedaner, T.; Korzun, V. Marker-assisted selection for disease resistance in wheat and barley breeding. *Phytopathology* **2012**, *102*, 560–566. [CrossRef] [PubMed]

18. Amarasinghe, C.C.; Simsek, S.; Brulè-Babel, A.; Fernando, W.G.D. Analysis of deoxynivalenol and deoxynivalenol-3-glucosides content in Canadian spring wheat cultivars inoculated with *Fusarium graminearum*. *Food Addit. Contam. A* **2016**, *33*, 1254–1264. [CrossRef] [PubMed]

19. Pestka, J. Toxicological mechanisms and potential health effects of deoxynivalenol and nivalenol. *World Mycotoxin J.* **2010**, *3*, 323–347. [CrossRef]

20. Knutsen, H.K.; Alexander, J.; Barregård, L.; Bignami, M.; Brüschweiler, B.; Ceccatelli, S.; Cottrill, B.; Dinovi, M.; Grasl-Kraupp, B.; Hogstrand, C.; et al. Risks to human and animal health related to the presence of deoxynivalenol and its acetylated and modified forms in food and feed. *EFSA J.* **2017**, *15*, 4718. [CrossRef]

21. European Commission. Commission Regulation (EC) No 1881/2006 of 19 December 2006 setting maximum levels for certain contaminants in foodstuffs. *Off. J. Eur. Union* **2006**, *364*, 5–24.

22. Busman, M.; Poling, S.M.; Maragos, C.M. Observation of T-2 toxin and HT-2 toxin glucosides from *Fusarium sporotrichioides* by liquid chromatography coupled to tandem mass spectrometry (LC-MS/MS). *Toxins* **2011**, *3*, 1554–1568. [CrossRef]

23. Nakagawa, H.; Ohmichi, K.; Sakamoto, S.; Sago, Y.; Kushiro, M.; Nagashima, H.; Yoshida, M.; Nakajima, T. Detection of a new *Fusarium* masked mycotoxin in wheat grain by high-resolution LC-Orbitrap MS. *Food Addit. Contam. A* **2011**, *28*, 1447–1456. [CrossRef]

24. Nakagawa, H.; Sakamoto, S.; Sago, Y.; Nagashima, H. Detection of type A trichothecene di-glucosides produced in corn by high-resolution liquid chromatography-Orbitrap mass spectrometry. *Toxins* **2013**, *5*, 590–604. [CrossRef]

25. Berthiller, F.; Dall'Asta, C.; Corradini, R.; Marchelli, R.; Sulyok, M.; Krska, R.; Adam, G.; Schuhmacher, R. Occurrence of deoxynivalenol and its 3-beta-D-glucoside in wheat and maize. *Food Addit. Contam. A* **2009**, *26*, 507–511. [CrossRef]

26. Dall'Asta, C.; Dall'Erta, A.; Mantovani, P.; Massi, A.; Galaverna, G. Occurrence of deoxynivalenol and deoxynivalenol-3-glucoside in durum wheat. *World Mycotoxin J.* **2013**, *6*, 83–91. [CrossRef]

27. Kostelanska, M.; Hajslova, J.; Zachariasova, M.; Malachova, A.; Kalachova, K.; Poustka, J.; Fiala, J.; Scott, P.M.; Berthiller, F.; Krska, R. Occurrence of deoxynivalenol and its major conjugate, deoxynivalenol-3-glucoside, in beer and some brewing intermediates. *J. Agric. Food Chem.* **2009**, *57*, 3187–3194. [CrossRef] [PubMed]

28. Lancova, K.; Hajslova, J.; Poustka, J.; Krplova, A.; Zachariasova, M.; Dostalek, P.; Sachambula, L. Transfer of *Fusarium* mycotoxins and "masked" deoxynivalenol (deoxynivalenol-3-glucoside) from field barley through malt to beer. *Food Addit. Contam. A* **2008**, *25*, 732–744. [CrossRef] [PubMed]

29. Maul, R.; Müller, C.; Rieß, S.; Koch, M.; Methner, F.J.; Nehls, I. Germination induces the glucosylation of the *Fusarium* mycotoxin deoxynivalenol in various grains. *Food Chem.* **2012**, *131*, 274–279. [CrossRef]

30. Spanic, V.; Zdunic, Z.; Drezner, G.; Sarkanj, B. The pressure of Fusarium disease and its relation with mycotoxins in the wheat grain and malt. *Toxins* **2019**, *11*, 198. [CrossRef] [PubMed]

31. MEBAK®. Methodensammlung der Mitteleuropäischen Analysenkommission. In *Raw Materials: Barley, Adjuncts, Malt, Hops and Hop Products*; Jacob, F., Ed.; Selbstverlag der MEBAK®: Freising-Weihenstephan, Germany, 2011.

32. Jin, Z.; Cao, Y.; Su, A.; Yu, Y.; Xu, M. Increase of deoxynivalenol during the malting of naturally *Fusarium* infected Chinese winter wheat. *Food Control.* **2018**, *87*, 88–93. [CrossRef]

33. Jin, Z.; Gillespie, J.; Barr, J.; Wiersma, J.J.; Sorrells, M.E.; Zwinger, S.; Gross, T.; Cumming, J.; Bergstrom, G.C.; Brueggeman, R.; et al. Malting of *Fusarium* Head Blight-infected rye (Secale cereale): Growth of *Fusarium graminearum*, trichothecene production, and the impact on malt quality. *Toxins* **2018**, *10*, 369. [CrossRef]

34. Jin, Z.; Zhou, B.; Gillespie, J.; Gross, T.; Barr, J.; Simsek, S.; Brueggeman, R.; Schwarz, P.B. Production of deoxynivalenol (DON) and DON-3-glucoside during the malting of *Fusarium* infected hard red spring wheat. *Food Control.* **2018**, *85*, 6–10. [CrossRef]

35. Yu, J.; Yin, H.; Dong, J.; Zhang, C.; Zhang, B.; Jin, Z.; Cao, Y. Pullulation of toxigenic *Fusarium* and deoxynivalenol in the malting of de minimis infected barley (*Hordeum vulgare*). *LWT Food Sci. Tech.* **2019**, *113*, 108242. [CrossRef]

36. Shellhammer, T.H. Beer Fermentations. In *The Oxford Handbook of Food Fermentations*; Bamforth, C.W., Ward, R.E., Eds.; Oxford University Press: Oxford, UK, 2014.

37. Carvalho, D.O.; Gonçalves, L.M.; Guido, L.F. Overall antioxidant properties of malt and how they are influenced by the individual constituents of barley and the malting process. *Compr. Rev. Food Sci. Food Saf.* **2016**, *15*, 927–943. [CrossRef]

38. Oliver, G. *The Oxford Companion to Beer*, 1st ed.; Oxford University Press: Oxford, UK, 2011.

39. Papadopoulou-Bouraoui, A.; Vrabcheva, T.; Valzacchi, S.; Stroka, J.; Anklam, E. Screening survey of deoxynivalenol in beer from the European market by an enzyme-linked immunosorbent assay. *Food Addit. Contam.* **2004**, *21*, 607–617. [CrossRef] [PubMed]

40. Zachariasova, M.; Hajslova, J.; Kostelanska, M.; Poustka, J.; Krplova, A.; Cuhra, P.; Hochel, I. Deoxynivalenol and its conjugates in beer: A critical assessment of data obtained by enzyme-linked immunosorbent assay and liquid chromatography coupled to tandem mass spectrometry. *Anal. Chim. Acta* **2008**, *625*, 77–86. [CrossRef] [PubMed]

41. Esslinger, H.M.; Narziss, L. Beer. In *Ullmann's Encyclopedia of Industrial Chemistry*; Wiley-VCH Verlag GmbH & Co. KGaA: Weinheim, Germany, 2009.

42. Bryła, M.; Ksieniewicz-Woźniak, E.; Waśkiewicz, A.; Szymczyk, K.; Jędrzejczak, R. Co-occurrence of nivalenol, deoxynivalenol and deoxynivalenol-3-glucoside in beer samples. *Food Control.* **2018**, *92*, 319–324. [CrossRef]

43. Varga, E.; Malachova, A.; Schwartz, H.; Krska, R.; Berthiller, F. Survey of deoxynivalenol and its conjugates deoxynivalenol-3-glucoside and 3-acetyl-deoxynivalenol in 374 beer samples. *Food Addit. Contam.* **2013**, *30*, 137–146. [CrossRef]

44. Malachova, A.; Varga, E.; Schwartz, H.; Krska, R.; Berthiller, F. Development, validation and application of an LC-MS/MS based method for the determination of deoxynivalenol and its conjugates in different types of beer. *World Mycotoxin J.* **2012**, *5*, 261–270. [CrossRef]

45. Bertuzzi, T.; Rastelli, S.; Mulazzi, A.; Donadini, G.; Amedeo, P. Mycotoxin occurrence in beer produced in several European countries. *Food Control.* **2011**, *22*, 2059–2064. [CrossRef]

46. Piacentini, K.C.; Savi, G.D.; Olivo, G.; Scussel, V.M. Quality and occurrence of deoxynivalenol and fumonisins in craft beer. *Food Control.* **2015**, *50*, 925–929. [CrossRef]

47. Roger, D.D. Deoxynivanol (DON) and fumonisins B1 (FB1) in artisanal sorghum opaque beer brewed in north Cameroon. *Afr. J. Microbiol. Res.* **2011**, *5*, 1565–1567. [CrossRef]

48. Kirui, M.C.; Alakonya, A.E.; Talam, K.K.; Tohru, G.; Bii, C.C. Total aflatoxin, fumonisin and deoxynivalenol contamination of busaa in Bomet county, Kenya. *Afr. J. Biotechnol.* **2014**, *13*, 2675–2678. [CrossRef]

49. Peters, J.; van Dam, R.; van Doorn, R.; Katerere, D.; Berthiller, F.; Haasnoot, W.; Nielen, M.W.F. Mycotoxin profiling of 1000 beer samples with a special focus on craft beer. *PLoS ONE* **2017**, *12*, e0185887. [CrossRef]

50. EFSA CONTAM Panel (EFSA Panel on Contaminants in the Food Chain). Scientific Opinion on risks for animal and public health related to the presence of nivalenol in food and feed. *EFSA J.* **2013**, *11*, 3262. [CrossRef]

51. Vidal, A.; Cano-Sancho, G.; Marin, S.; Ramos, A.J.; Sanchis, V. Multidetection of urinary ochratoxin A, deoxynivalenol and its metabolites: Pilot time-course study and risk assessment in Catalonia, Spain. *World Mycotoxin J.* **2016**, *9*, 597–612. [CrossRef]

52. Heyndrickx, E.; Sioen, I.; Huybrechts, B.; Callebaut, A.; De Henauw, S.; De Saeger, S. Human biomonitoring of multiple mycotoxins in the Belgian population: Results of the BIOMYCO study. *Environ. Int.* **2015**, *84*, 82–89. [CrossRef] [PubMed]

53. Warth, B.; Sulyok, M.; Fruhmann, P.; Berthiller, F.; Schuhmacher, R.; Hametner, C.; Adam, G.; Frohlich, J.; Krska, R. Assessment of human deoxynivalenol exposure using an LC-MS/MS based biomarker method. *Toxicol. Lett.* **2012**, *211*, 85–90. [CrossRef] [PubMed]

54. Grajewski, J.; Kosicki, R.; Twarużek, M.; Błajet-Kosicka, A. Occurrence and Risk Assessment of Mycotoxins through Polish Beer Consumption. *Toxins* **2019**, *11*, 254. [CrossRef]

55. Pascari, X.; Ortiz-Solá, J.; Marín, S.; Ramos, A.J.; Sanchis, V. Survey of mycotoxins in beer and exposure assessment through the consumption of commercially available beer in Lleida, Spain. *LWT Food Sci. Technol.* **2018**, *92*, 87–91. [CrossRef]

56. Rodriguez-Carrasco, Y.; Fattore, M.; Albrizio, S.; Berrada, H.; Manes, J. Occurrence of *Fusarium* mycotoxins and their dietary intake through beer consumption by the European population. *Food Chem.* **2015**, *178*, 149–155. [CrossRef]

57. European Food Safety Authority. Deoxynivalenol in food and feed: Occurrence and exposure. *EFSA J.* **2013**, *11*, 3379. [CrossRef]

58. Yoshinari, T.; Sakuda, S.; Furihata, K.; Furusawa, H.; Ohnishi, T.; Sugita-Konishi, Y.; Ishizaki, N.; Terajima, J. Structural determination of a nivalenol glucoside and development of an analytical method for the simultaneous determination of nivalenol and deoxynivalenol, and their glucosides, in wheat. *J. Agric. Food Chem.* **2014**, *62*, 1174–1180. [CrossRef]

59. European Commission. Commission Regulation (EC) No 519/2014 of 16 May 2014 amending Regulation (EC) No 401/2006 as regards methods of sampling of large lots, spices and food supplements, performance criteria for T-2, HT-2 toxin and citrinin and screening methods of analysis. *Off. J. Eur. Union* **2014**, *147*, 29–43.

60. Bryła, M.; Ksieniewicz-Woźniak, E.; Yoshinari, T.; Waśkiewicz, A.; Szymczyk, K. Contamination of wheat cultivated in various regions of Poland during 2017 and 2018 agricultural seasons with selected trichothecenes and their modified forms. *Toxins* **2019**, *11*, 88. [CrossRef] [PubMed]

Article

Acrylamide Reduction Strategy in Combination with Deoxynivalenol Mitigation in Industrial Biscuits Production

Michele Suman [1,*], **Silvia Generotti** [1,2], **Martina Cirlini** [2] and **Chiara Dall'Asta** [2]

[1]   Advanced Research Laboratory, Barilla G. e R. Fratelli S.p.A., Via Mantova, 166-43122 Parma, Italy
[2]   Department of Food & Drug, University of Parma, Parco Area delle Scienze, 95/A-43124 Parma, Italy
*   Correspondence: michele.suman@barilla.com

Received: 12 August 2019; Accepted: 22 August 2019; Published: 27 August 2019

Abstract: Acrylamide is formed during baking in some frequently consumed food products. It is proven to be carcinogenic in rodents and a probable human carcinogen. Thus, the food industry is working to find solutions to minimize its formation during processing. To better understand the sources of its formation, the present study is aimed at investigating how acrylamide concentration may be influenced by bakery-making parameters within a parallel strategy of mycotoxin mitigation (focusing specifically on deoxynivalenol—DON) related to wholegrain and cocoa biscuit production. Among Fusarium toxins, DON is considered the most important contaminant in wheat and related bakery products, such as biscuits, due to its widespread occurrence. Exploiting the power of a Design of Experiments (DoE), several conditions were varied as mycotoxin contamination levels of the raw materials, recipe formulation, pH value of dough, and baking time/temperature; each selected treatment was varied within a defined range according to the technological requirements to obtain an appreciable product for consumers. Experiments were performed in a pilot-plant scale in order to simulate an industrial production and samples were extracted and analysed by HPLC-MS/MS system. Applying a baking temperature of 200 °C at the highest sugar dose, acrylamide increased its concentration, and in particular, levels ranged from 306 ± 16 µg/Kg d.m. and 400 ± 27 µg/Kg d.m. in biscuits made without and with the addition of cocoa, respectively. Conversely, using a baking temperature of 180 °C in the same conditions (pH, baking time, and sugar concentrations), acrylamide values remained below 125 ± 14 µg/Kg d.m. and 156 ± 15 µg/Kg d.m. in the two final products. The developed predictive model suggested how some parameters can concretely contribute to limit acrylamide formation in the final product, highlighting a significant role of pH value (correlated also to sodium bicarbonate raising agent), followed by baking time/temperature parameters. In particular, the increasing range of baking conditions influenced in a limited way the final acrylamide content within the parallel effective range of DON reduction. The study represents a concrete example of how the control and optimization of selected operative parameters may lead to multiple mitigation of specific natural/process contaminants in the final food products, though still remaining in the sensorial satisfactory range.

Keywords: acrylamide; deoxynivalenol; multiple mitigation strategies; design of experiments; bakery food processing; biscuits

Key Contribution: The study here reported represents a concrete example of how the control and optimization of selected operative industrial parameters may lead to multiple mitigation of specific natural/process contaminants in the final food products: in this specific case Deoxynivalenol & Acrylamide within the contest of biscuits production.

---

1. Introduction

Process-related mitigation strategies are among the most promising tools for controlling and minimizing mycotoxins, i.e., Fusarium mycotoxins, in cereal-based products [1]. Among Fusarium mycotoxins, deoxynivalenol (DON), along with its modified forms, is considered the most important contaminant in wheat and related bakery products, such as biscuits, due to its widespread occurrence [2]. It has been shown that milling and baking/roasting may effectively reduce the amount of DON in the bakery. At this purpose, Generotti et al. reported on the strategic mitigation of DON and its related compounds deoxynivalenol-3-glucoside (DON3Glc) and culmorin, during two different biscuit processes, exploiting the synergistic effects of recipe formulation and thermal treatment [3]. A significant reduction was achieved while not negatively impacting product quality and identity. However, increased time and temperature during baking might favor the formation of process-derived contaminants, such as acrylamide, in the final product.

Besides often reported among the most studied contaminants in cereal-based products, mycotoxins and acrylamide are almost unavoidable compounds, being the former natural toxins accumulated in crops under natural field conditions, and the latter process-related compounds formed in food during manufacturing, as a consequence of chemical reaction triggered by processing.

Although their toxic effects have been deeply studied as single compounds, little to nothing is known about combined effects exerted in animals and humans. Nonetheless, the scientific community is posing increasing emphasis on health concern related to the exposure to chemical mixtures, as recently reported by the European Food Safety Authority [4]. Therefore, the identification of possible strategies for the simultaneous mitigation of mycotoxins and acrylamide, are of upmost interest for the agro-food sector.

Concerning process-related compounds, food industry and the European Commission have undertaken extensive efforts since 2002, when scientists from the Swedish National Food Authority and the University of Stockholm reported high levels of acrylamide in normally cooked starch-rich food (compared to what had been reported earlier in other food commodities), in order to investigate pathways of formation and to reduce its levels in processed food.

Acrylamide is typically formed from an amino acid, primarily asparagine, and a reducing sugar such as fructose or glucose in starchy food products during high temperature cooking, including frying, baking and roasting through a series of reactions, known as Maillard reactions. Its formation starts at temperatures around 120 °C and peaks at temperatures between 160 and 180 °C [5,6]. Due to its toxicity and possible carcinogenic effects to humans (IARC—International Agency for Research on Cancer [7]), the European Commission has established mitigation measures and benchmark levels for its concrete reduction in food [8,9].

Several attempts have been made so far to develop mitigation strategies for acrylamide formation in bakery products, mainly focused on the processing stage and recipes [10–17]. Each strategy could present limiting factors in their applicability depending on the product type and industrial settings as feasibility and compatibility with processing, formulation, impact on sensory and nutritional characteristics, regulatory compliance and costs.

Acrylamide formation is favored by a high baking temperature and time treatment [17–23], therefore a decreasing of thermal input represents an effective way of mitigation. Reduction can be obtained by applying prolonged heating at lower temperatures, or at lower pressure than the atmospheric one [16] or by optimizing the oven temperature profile. On the other hand, a decreased thermal input may significantly affect the achievement of both appropriate hygienic properties and sensorial acceptance of the final product.

Moving from our previous studies [24–27], the present investigation is aimed at verifying how acrylamide concentration in bakery products, such as wholegrain and cocoa biscuits, is affected by modifications of technological parameters (recipe formulation and baking time/temperature) during biscuit-making process (Figure 1), while at the same targeting potential DON reduction and without affecting the sensory properties.

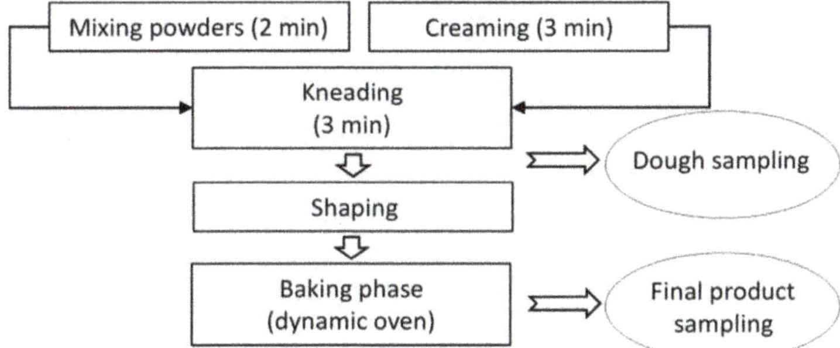

Figure 1. Scheme of wholegrain and cocoa biscuit production.

In this regard, predictive models would represent a time and cost-saving tool for finding the most suitable conditions for minimizing both natural occurring and process-related contaminants. In the present work, we have exploited them to estimate acrylamide evolution in a considered system and to manage the industrial process to optimize the role of involved parameters with regard to its formation.

Starting from naturally DON contaminated raw material, the experiments were performed using statistical Design of Experiment (DoE) schemes to explore the relationship between the analytical responses and independent variables conducting to an overall optimization of the baking process [28]. In particular, this approach was conducted in order to consider only those modifications that can be really applied to the industrial scale, obtaining a final product appreciable by consumers and remaining in an acceptable technological range.

2. Results and Discussion

The present study was carried out to better understand whether and how acrylamide evolution could be influenced by selected technological factors within a strategy of DON mitigation related to wholegrain and cocoa biscuit production. Different parameters were considered and a variation range was defined, as shown in Table 1. The statistical model required 19 single experiments per technological process; acrylamide level was measured by LC-MS/MS according to a previously published method with slight modifications [29].

Table 1. Wholegrain and cocoa biscuit experiments—pilot plant processing conditions.

TREATMENT	WHOLEGRAIN BISCUIT MAKING			COCOA BISCUIT MAKING		
	Minimum	Central Point	Maximum	Minimum	Central Point	Maximum
DON bran level (μg/kg)	600	1050	1500	600	1050	1500
Dextrose (%)	15	19	23	15	19	23
Milk (%)	-	-	-	5	6.5	8
Eggs (%)	4	6	7	-	-	-
Margarine (%)	10	15	20	-	-	-
pH value	5	6.5	8	5	6.5	8
Baking time (min)	5	6.5	8	5	6.5	8
Baking temperature (°C)	180	190	200	180	190	200

2.1. Statistical Elaboration of the Experimental Model

At the end of the analysis performed on the collected samples, MODDE software extracted an answer concerning robustness and prediction capability of the method, evaluating model efficiency by fitting (R2) and prediction (Q2) values. Replicate values of acrylamide reduction per experiment were expressed on dry matter basis and averaged for the statistical elaboration. Partial least-squares (PLS) was chosen as the statistical regression treatment. The two statistical models (wholegrain and cocoa biscuit model) gave high fitting ($R2 > 0.7$) and good prediction value ($Q2 \geq 0.5$), referring to the MODDE software output settings. Model robustness was also confirmed by ANOVA plot (Figure 2), being standard deviation of the regression much larger than standard deviation of the residuals.

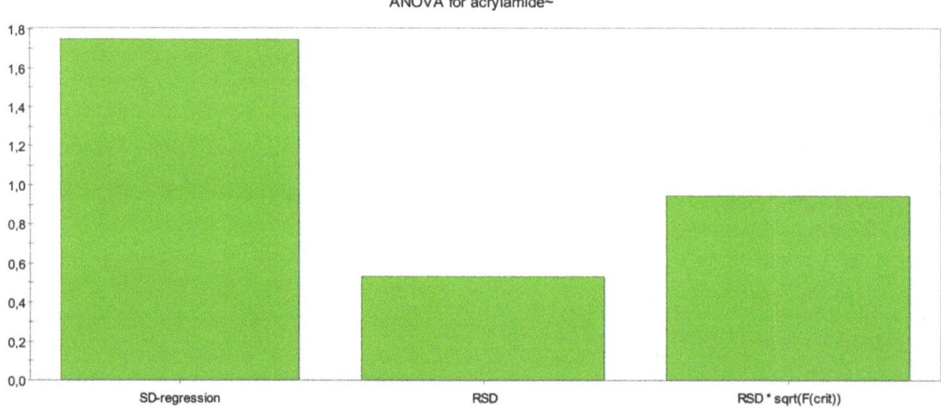

Figure 2. Design of Experiments on acrylamide levels within cocoa biscuit-making—Model Robustness by ANOVA plot.

2.2. Acrylamide Evolution within Biscuit-Making Technology

All values were collected in a Variable Importance Plot (VIP, Figure 3) that enables (and illustrates in an effective and condensed way) an understanding of the effect of each factor in terms of influence on acrylamide response.

In the present case, VIP would suggest that the pH value, has the most relevant effect on the final acrylamide level: in fact pH increase is responsible for an acceleration of the reaction between asparagine and the reducing sugars, followed by baking time/temperature parameters.

With regard to the other minor ingredients, dextrose (or glucose) content confirm (as reported in previous scientific literature findings) to contribute to the overall acrylamide increase. When high dextrose level and higher thermal input are employed (200 °C for 8 min) an acrylamide increase up to 120% was observed (data not shown) with respect to the central point. On the other hand, a combination of lower dextrose content and moderate thermal input (180 °C for 8 min) may lead to a reduction up to 77%.

This is consistent with literature studies in which it was demonstrated that increasing the quantity of sugars in cookies formulation, as glucose and sucrose, the concentration of acrylamide raised up especially when a temperature of 205° was applied for 11 min. Moreover, the choice of the sugar resulted crucial for keeping acrylamide formation under control. Using glucose, a reducing compound, and applying the same baking conditions, acrylamide content increased more in respect to sucrose. So, the authors suggested to use sucrose in order to obtain a reduction of about the 50% of acrylamide production [30,31]. Similar results were showed by Vass et al. [32] whom replaced invert sugar syrup with sucrose in wheat crackers, obtaining an acrylamide reduction by 60%. In addition, the same studies speculated that applying baking temperatures of 160°C the formation of acrylamide remains below 150 µg/Kg [30]. Also in this study, the measured amount of acrylamide presented values below or close to 150 µg/Kg when the baking temperature did not exceed 180 °C (Table 2).

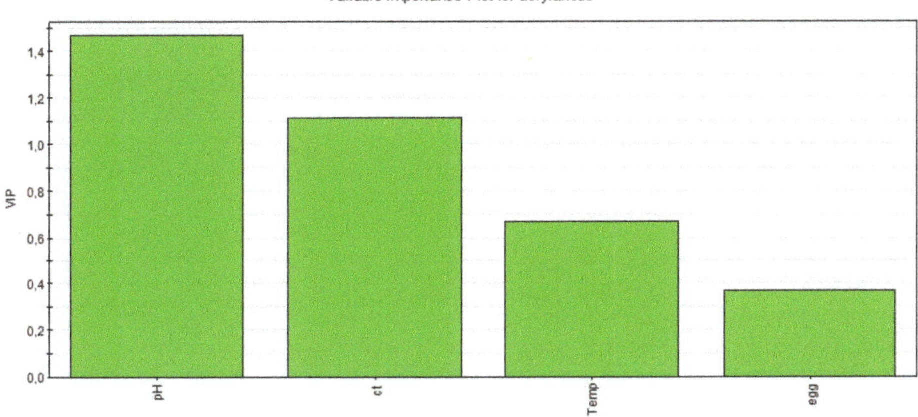

Figure 3. Variable Importance Plot (VIP) obtained for the data referred to cocoa biscuit-making process: influence of each main factor with respect to acrylamide response: pH, ct (cooking time), Temperature (Temp), Egg.

Table 2. Analytical results for acrylamide levels throughout several wholegrain and cocoa biscuit-making process trials.

Sample	Baking Stage					
	DON in Wholegrain Flour (µg/kg d.m.) *	NaHCO3 (g)	Time (min)	Temperature (°C)	Acrylamide (µg/kg d.m.) *	DON (µg/kg d.m.) *§
Wholegrain biscuits	219 ± 8	8	5	180	16 ± 3	154 ± 1
	304 ± 9	8	8	180	125 ± 14	192 ± 4
	219 ± 8	8	5	200	66 ± 10	188 ± 7
	304 ± 9	8	8	200	306 ± 16	189 ± 7
Cocoa biscuits	219 ± 8	8	5	180	43 ± 13	129 ± 7
	304 ± 9	8	8	180	156 ± 15	208 ± 2
	219 ± 8	8	5	200	185 ± 15	154 ± 1
	304 ± 9	8	8	200	400 ± 27	192 ± 4

* Data expressed as mean value ± standard deviation. § Data reported and discussed elsewhere [3].

Furthermore, some authors showed how acrylamide mitigation can be achieved by adding amino acids or protein-based ingredients to food, which may influence the reaction pathway or favor AA degradation [33–35]. In our study, no significant mitigation was obtained probably due to the close variation range related to milk/egg content in the recipe formulation.

Basically, acrylamide content seems to be affected by the food matrix, being higher for the cocoa biscuits than for the wholegrain biscuits. This could be related to the additional acrylamide load related to cocoa beans roasting (Table 2). It was indeed demonstrated that acrylamide could be present in roasted cocoa beans in the order of mg/Kg and the content may depend form the temperature used during roasting [36].

With regard to the sodium bicarbonate content and the correspondent pH variation, a potential reduction in acrylamide level higher than 50% could be achieved in the finished product, remaining within an acceptable range from the sensorial point of view. Sodium bicarbonate could indeed limit the formation of acrylamide if compared to other raising agents as ammonium hydrogen carbonate. This was demonstrated during experiments conducted on biscuits, in which also the addition of tartaric and citric acids was tested in order to reduce acrylamide content. The authors showed that a reduction of about 70% of acrylamide content was achieved when sodium bicarbonate was applied [31].

Taking into account the previously mentioned mycotoxin mitigation strategy, baking time/temperature play an important role in order to achieve a parallel significant DON mitigation; considering experiments carried out within the most severe time/temperature baking conditions, the greatest effect was observed with the baking step being performed at 200 °C for 8 min. In particular, an increase in time during the baking phase, in an acceptable technological range, can effectively reduce DON content in the final product [3].

Notably, an acrylamide reduction ranged from 77% to 100% was achieved in the finished product when baking was conducted at 180 °C for 5 min, though still remaining in the sensorial satisfactory range.

Overall, data collected within this study allow to proper set the baking time and temperature for controlling mycotoxin and acrylamide content in the finished product, as suggested by the response Contour Plot (Figure 4). The increase of baking parameters (which also goes in the direction of the obvious industrial requirement to speed-up the process, in order to increase the productivity as much as possible) within a range of mycotoxin mitigation (up to 20% of reduction) affects in a restricted manner the final acrylamide content (difference of the values of acrylamide expressed as predicted increase, named "delta acrylamide" in the figure caption), without implications on the organoleptic properties and consumer safety.

As a major outcome of this study, this allows one to design proper synergistic mitigation strategies for multiple contaminants along the food production chain.

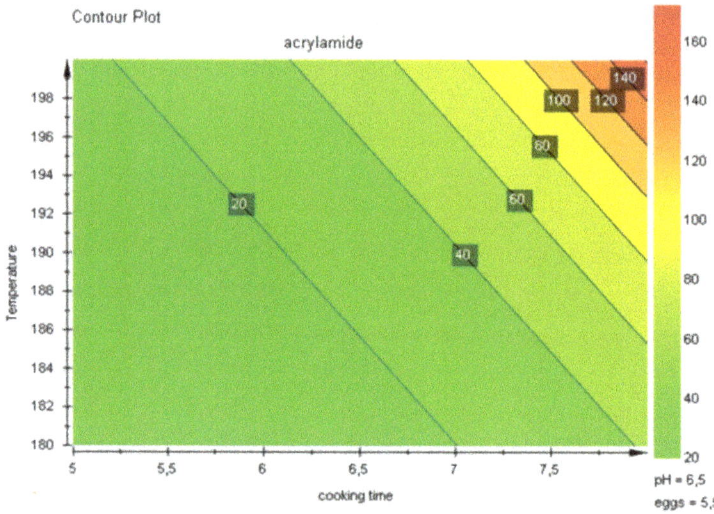

Figure 4. Contour Plot: delta-acrylamide (predicted increase) values of the cocoa biscuit-making experiments; Temperature vs. cooking time.

3. Materials and Methods

3.1. Chemicals

Methanol and formic acid (p.a.), both HPLC gradient grade, were obtained from BDH VWR International Ltd. (Poole, UK). Acetonitrile was purchased from J.T. Baker (Deventer, The Netherlands) and ammonium acetate (MS grade) and glacial acetic acid (p.a.) were obtained from Sigma-Aldrich (Vienna, Austria). Standard acrylamide solution was purchased from Sigma-Aldrich (Milan, Italy). Acrylamide internal standard (13C3-acrylamide, 1 mg/mL in methanol) was obtained from Cambridge Isotope Laboratories, Inc. (Andover, MA, USA). Deionized water was used for all procedures. Water was purified successively by reverse osmosis and a Milli-Q plus system from Millipore (Molsheim, France). Deoxynivalenol standard was obtained from RomerLabs®Inc. (Tulln, Austria). OASIS® HLB 3 cc (60 mg) extraction cartridges were purchased from Waters (Manchester, UK). Glass vials with septum screw caps were purchased from Phenomenex (Torrance, CA, USA). Centrifugal filter units (Ultrafree MC 0.22 mm, diameter 10 mm) were obtained from Millipore (Billerica, MA, USA).

3.2. Biscuit-Making Production in Pilot-Plant Scale

Pilot-plant scale experiments were performed according to a previously published study [3]. Briefly, three batches of wheat bran naturally infected with Fusarium spp. were analysed with a focus on DON, selected and mixed with a blank wheat flour for the wholegrain biscuit production. Concerning cocoa biscuit trials, the same three mix flours were employed and three different batches of cocoa powder were selected.

Different doughs were prepared in order to obtain a final dough of about 1000 ± 30 g; regarding wholegrain biscuits, the ingredients were wheat flour (60%), bran (7%), cream of tartar, glucose syrup, and salt. Eggs, margarine, and dextrose were added depending on the value reported in recipes obtained from the experimental design (Table 3). Sodium bicarbonate was added in order to reach the appropriate pH value. Water amount ranged from 1.6% to 5%, depending on the technological requirements.

Table 3. Experimental data set for screening variables effects on acrylamide levels within the wholegrain biscuit-making process steps: full-factorial central composite design.

Experiment Number	DON in Bran (µg/kgd.m.) [1]	Dextrose (%)	Margarine (%)	pH	Eggs (%)	Baking Stage	
						Time (min)	Temperature (°C)
1	600 ± 16	15	10	5	4	5	180
2	1500 ± 92	15	10	5	7	5	200
3	600 ± 16	23	10	5	7	8	180
4	1500 ± 92	23	10	5	4	8	200
5	600 ± 16	15	20	5	7	8	200
6	1500 ± 92	15	20	5	4	8	180
7	600 ± 16	23	20	5	4	5	200
8	1500 ± 92	23	20	5	7	5	180
9	600 ± 16	15	10	8	4	8	200
10	1500 ± 92	15	10	8	7	8	180
11	600 ± 16	23	10	8	7	5	200
12	1500 ± 92	23	10	8	4	5	180
13	600 ± 16	15	20	8	7	5	180
14	1500 ± 92	15	20	8	4	5	200
15	600 ± 16	23	20	8	4	8	180
16	1500 ± 92	23	20	8	7	8	200
17	1050 ± 48	19	15	6.5	6	6.5	190
18	1050 ± 48	19	15	6.5	6	6.5	190
19	1050 ± 48	19	15	6.5	6	6.5	190

[1] Data expressed as mean value ± standard deviation.

In order to produce cocoa biscuits, 45% of wheat flour, 7% of bran, 4% of cocoa powder, margarine, glucose syrup, and salt were employed. Milk and dextrose amount were indicated in the model generated by Design of Experiment (Table 4). Sodium bicarbonate was added in order to reach the appropriate pH value. The optimal amount of water to be added to each dough sample was established on the basis of internal technological knowledge.

Table 4. Experimental data set for screening variables effects on acrylamide levels within the cocoa biscuit-making process steps: full-factorial central composite design.

Experiment Number	DON in Bran (μg/kgd.m.) [1]	Dextrose (%)	Milk (%)	pH	Baking Stage	
					Time (min)	Temperature (°C)
1	600 ± 16	23	5	5	5	180
2	600 ± 16	15	8	8	5	180
3	1500 ± 92	15	5	5	8	180
4	1500 ± 92	23	8	8	8	180
5	600 ± 16	23	5	8	5	200
6	1500 ± 92	23	8	5	5	200
7	1500 ± 92	15	5	8	8	200
8	600 ± 16	15	8	5	8	200
9	1500 ± 92	23	5	8	5	180
10	600 ± 16	23	8	5	5	180
11	600 ± 16	15	5	8	8	180
12	1500 ± 92	15	8	5	8	180
13	600 ± 16	15	5	5	5	200
14	1500 ± 92	15	8	8	5	200
15	1500 ± 92	23	5	5	8	200
16	600 ± 16	23	8	8	8	200
17	1050 ± 48	19	6.5	6.5	6.5	190
18	1050 ± 48	19	6.5	6.5	6.5	190
19	1050 ± 48	19	6.5	6.5	6.5	190

[1] Data expressed as mean value ± standard deviation.

The process for wholegrain and cocoa biscuit production consisted essentially of the following steps: creaming, dough preparation, and baking step. Firstly, wheat flour was mixed with all solid powder ingredients using a test planetary kneader for 2 min. Dextrose and margarine were mixed separately by using another test planetary kneader for 3 min (creaming step). At a later stage, cream and powders were mixed together for 3 min. Dough was shaped and rounded pieces of about 4 cm diameter (approximately 10 g) were obtained from dough and rested for 10 min at room temperature. Baking step was performed in a pilot-scale dynamic oven (Tagliavini, Parma, Italy). The overall process is summarized in Figure 1. Nineteen different tests for each process were performed (Tables 3 and 4, respectively).

Acrylamide content was examined in mix powders, before and after baking process. Before the acrylamide content analysis, samples were stored at −20 °C. Each sample was extracted and analyzed in duplicate.

3.3. Moisture Content Determination

The moisture contents of mix flours, doughs, and baked products were measured by taking a 5 g ground sample and heating it in a thermostatic oven at 105 °C for 6 h. All the results were compared on a dry matter (d.m.) basis.

3.4. Experimental Design and Statistical Evaluation

Design of Experiments (DoE) is used in many industrial issues, in the development and optimization of manufacturing processes, making a set of experiments representative with regards to a given question. DoE is a series of tests in which purposeful changes are made to the input variables of a system or process and the effects on response variables are measured: the analyst is interested in studying the synergistic effects of some interventions (the "treatments") to optimize the final process [28].

Among the wholegrain and cocoa biscuit-making parameters, several conditions were varied during the experiments: DON contamination level on wheat bran; dextrose, margarine, egg and milk content (as percentage in recipes); pH value (as sodium bicarbonate content) and baking time and temperature. Each selected treatment was varied within a range defined according to the technological requirements to obtain an organoleptically appreciable product for consumers: the central experimental values, indicated in Table 1, represent the optimal combination of ingredients/recipe and operative conditions that permit to achieve the most appropriate finished product.

Experimental data were then analysed by a multi-variate analysis approach based on the partial least-squares (PLS) technique, using a dedicated statistical package (MODDE software, version 9.1, 2012; Umetrics, Umea, Sweden).

3.5. Sample Extraction and Instrumental Conditions—Deoxynivalenol

Concerning Deoxynivalenol, samples were extracted according to a previously published procedure [3,25] with slight modifications. Briefly, a total of 10.00 g of flour or dough or biscuit sample were extracted with 100 mL of an acetonitrile/water (84:16, *v/v*) mixture by homogenization at a medium-to-high speed for 2 min using a mixer (Oster, New York, USA). The extract was allowed to settle for 15 min. Afterwards, 5 mL were poured into a 10 mL vial, and evaporated to dryness under a nitrogen stream. The extract was reconstituted with 100 mL of 13C-DON internal standard solution (100 ng/mL in methanol) and 900 mL of water. Each extraction cartridge column was activated using 2 mL of methanol, and 2 mL of methanol:water (10:90, *v/v*). The sample extract was then slowly passed through the OASIS® HLB 3 cc (60 mg) (Waters, Manchester, UK) column using a vacuum chamber system. A solution of methanol:water (20:80, *v/v*) was used for washing, followed by elution with 1 mL of methanol. The eluate was evaporated under a gentle stream of nitrogen, and the residue was dissolved in 200 mL of eluent A (methanol:water, 20:80 *v/v*, 0.5% acetic acid, and 1 mM ammonium acetate) prior to UHPLCMS/MS analysis. Ultrahigh-performance liquid chromatography (UHPLC) was performed using a Dionex Ultimate® 3000 LC systems (Thermo Fisher Scientific Inc., Waltham, MA, USA) and a Kinetex Biphenyl column (2.6 mm; 100 × 2.10 mm; Phenomenex). The flow rate of the mobile phase was 400 mL/min, and the injection volume was 20 mL. The column oven was set to 30 °C. A linear binary gradient composed of (A) water (0.5% acetic acid, 1 mM ammonium acetate) and (B) methanol (0.5% acetic acid, 1 mM ammonium acetate) was employed. The gradient was as follows: 0–4 min to 40% B; 4–20 min to 80% B; 20–22 min, isocratic step 80% B; finally, a re-equilibration step at 10% B (the initial value) was performed for another 3 min, bringing the total analysis time to 25 min. Before UHPLC-MS/MS analysis, all samples were filtered through centrifugal filter units for clarification. ESI-MS/MS was carried out by a Q-Exactive (Thermo Fisher Scientific Inc., Waltham, MA, USA) mass spectrometer. Experiments were performed in full MS data scan for quantification and data-dependent scan with the following settings: the capillary temperature was set to 300 °C; the sheath gas and auxiliary gas flow rates were set to 40 and 10 units, respectively; the spray voltage was set to 3500 kV; the S-lens RF level was set to 55 V. All equipment control and data processing were performed by Excalibur software (Thermo Fisher Scientific Inc., Waltham, MA, USA). Deoxynivalenol measurements in all the samples were performed using isotopically labeled standard and calibration vs. matrix-matched standards.

3.6. Sample Extraction and Instrumental Conditions—Acrylamide

Sample extraction for acrylamide was performed according to a previously published procedure [29] with slight modifications. Briefly, samples were finely ground in a blender to homogeneity before extraction. 1 g of sample was weighed into a polypropylene graduated conical tube and different volumes of a 300 μl ml-1 internal standard solution (13C3-labeled acrylamide in 0.1% (*v/v*) formic acid) followed by 10 mL 0.1% (*v/v*) formic acid were added on the base of the acrylamide concentrations supposed to be present in the samples. After mixing for 10 min on a vortexer the extract was centrifuged at 1,0000 rpm for 5 min. A 3-mL portion of clarified solution was removed avoiding to collect top oil layer when present and filtered through a 0.45 μm nylon syringe filter (Phenomenex, Torrance, CA, USA) before injection into the HPLC-MS/MS system (injection volume 10 μL). LC-ESI-MS/MS in positive ion mode analysis was achieved using a Surveyor LC quaternary pump separation system (Thermo Fisher Scientific Inc.) coupled to a linear ion trap LXQ mass spectrometer (Thermo Fisher Scientific Inc., Waltham, MA, USA).

Chromatographic separation was performed using a Synergi Hydro-RP (150 × 2.0 mm) 4 μm analytical column (Phenomenex, Torrance, CA, USA). Elution was carried out at a flow rate of 0.2 mL/min, in isocratic conditions, at 30 °C using as mobile phase a mixture of 98.9% water, 1% methanol and a.1% formic acid (*v/v/v*). In these conditions, the retention time of acrylamide was about 4 min. A time programmed valve was used to discard the eluate from the column for the first 2.5 min in order to eliminate the compounds with retention times shorter than acrylamide. At 8 min the column flow was again diverted and the mobile phase changed to 100% methanol in order to clean the column from strongly retained compounds within a total run time of 10 min. MS/MS conditions were set as follows: capillary temperature was set to 160 °C; the sheath gas was set to 35 units; the spray voltage was kept at 4500 V; capillary voltage was kept at 9 V. All parts of the equipment and data processing were performed by the computer software Xcalibur (Thermo Fisher Scientific Inc.). MS/MS analysis was carried out by selecting the ions at *m/z* 72 and *m/z* 75 as precursor ions for acrylamide and 13C3-acrylamide respectively.

The area of the chromatographic peaks of the extracted ion at *m/z* 55, due to the transition 72 > 55, and at *m/z* 58, due to the transition 75 > 58 were used for the quantitative analysis. The quantitative analysis was carried out with the method of the internal standard. The relative response factor of acrylamide with respect to 13C3-acrylamide was calculated daily by analysing a standard solution.

4. Conclusions and Outlook

Since precursors of acrylamide are present in the dough, modifications in recipe formulation and time-temperature control during baking process could be actually used to reduce acrylamide content in biscuits.

In the present study, the influence exerted by modifying ingredients and industrial conditions on acrylamide levels within a parallel mycotoxin mitigation strategy was investigated.

The obtained processing models showed a good fitting, robustness, and prediction capability, suggesting the most significant parameters. These can concretely contribute to the reduction of acrylamide levels in the final food product.

Acrylamide formation is evidently baking time- and temperature-dependent, therefore prolongation of heat treatments results in higher contents of acrylamide; however, when such parameters are moved within the optimal range for DON mitigation, the actual increase affects, in a limited way, the final acrylamide content without significant implications on the organoleptic properties.

In conclusion, the present report demonstrates the effectiveness of a careful design of process parameters for the mitigation of multiple contaminants in the final product, thus remaining within the consumer's sensorial acceptance.

Author Contributions: Conceptualization, M.S. and C.D.; Data Elaboration and Validation, M.S., S.G., C.D. and M.C.; Pilot Plant Trials, S.G.; Formal analysis, S.G. and M.C.; Writing—original draft, S.G. and M.S.; Writing—review & editing, M.S., C.D. and M.C.

Funding: This research received no external funding.

Acknowledgments: The authors would like to thank Claudio Dall'Aglio (Barilla Pilot Plants) for his collaboration in the pilot plant trials, Nadia Morbarigazzi (Barilla Research & Development) for her suggestions and fruitful discussions and Dante Catellani (Barilla Food Research Labs) for his time and availability.

Conflicts of Interest: The authors declare no conflict of interest.

References

1. Karlovsky, P.; Suman, M.; Berthiller, F.; De Meester, J.; Eisenbrand, G.; Perrin, I.; Oswald, I.P.; Speijers, G.; Chiodini, A.; Recker, T.; et al. Impact of food processing and detoxification treatments on mycotoxin contamination. *Mycotoxin Res.* **2016**, *32*, 179–205. [CrossRef] [PubMed]
2. European Food Safety Authority. Deoxynivalenol in food and feed: Occurrence and exposure. *EFSA J.* **2013**, *11*, 3379–3435.
3. Generotti, S.; Cirlini, M.; Šarkanj, B.; Sulyok, M.; Berthiller, F.; Dall'Asta, C.; Suman, M. Formulation and processing factors affecting trichothecene mycotoxins within industrial biscuit-making. *Food Chem.* **2017**, *229*, 597–603. [CrossRef] [PubMed]
4. European Food Safety Authority. Guidance on harmonised methodologies for human health, animal health and ecological risk assessment of combined exposure to multiple chemicals. *EFSA J.* **2019**, *17*, 5634–5711.
5. Mottram, D.S.; Wedzicha, B.L.; Dodson, A.T. Acrylamide is formed in the Maillard reaction. *Nature* **2004**, *419*, 448–449. [CrossRef] [PubMed]
6. Törnqvist, M. Acrylamide in food: The discovery and its implication. In *M. Friedman & D. Mottram, Chemistry and Safety of Acrylamide in Food*; Springer Science: New York, NY, USA; Business Media, Inc.: New York, NY, USA, 2005; pp. 1–9.
7. IARC (International Agency for Research on Cancer). *IARC Monographs on the Evaluation of Carcinogenic Risk to Humans: Some Industrial Chemicals*; pp. 39, 41 (1986); Suppl. 7, pp. 56 (1987); pp. 60, 389 (1994); IARC: Lyon, France, 1994; p. 60.
8. European Commission. Commission Recommendation 2013/647/EU of 8 November 2013 on investigations into the levels of acrylamide in food. *Off. J. Eur. Union* **2013**, *301*, 15–17.
9. Draft Commission Regulation (EU). Commission Reg. (EU) on the Application of Control & Mitigation Measures to Reduce the Presence of Acrylamide in Food Ref. Ares (2017)2895100-09/06/2017. Available online: https://ec.europa.eu/info/law/better-regulation/initiatives/ares-2017-2895100_en (accessed on 1 June 2019).
10. Claus, A.; Carle, R.; Schieber, A. Acrylamide in cereal products: A review. *J. Cereal Sci.* **2008**, *47*, 118–133. [CrossRef]
11. Konings, E.J.M.; Ashby, P.; Hamlet, C.G.; Thompson, G.A.K. Acrylamide in cereal and cereal products: A review on progress in level reduction. *Food Addit. Contam.* **2007**, *24*, 47–59. [CrossRef]
12. Medeiros Vinci, R.; Mestdagh, F.; De Meulenaer, B. Acrylamide formation in fried potato products-present and future, a critical review on mitigation strategies. *Food Chem.* **2012**, *133*, 1138–1154. [CrossRef]
13. Sadd, P.A.; Hamlet, C.G.; Liang, L. Effectiveness of methods for reducing acrylamide in bakery products. *J. Agric. Food Chem.* **2008**, *50*, 4998–5006. [CrossRef]
14. Stadler, R.H. The formation of acrylamide in cereal products and coffee. In *Acrylamide and Other Hazardous Compounds in Heat-Treated Foods*; Skog, K., Alexander, J., Eds.; Woodhead Publishing: Cambridge, UK, 2006; pp. 23–40.
15. Anese, M.; Suman, M.; Nicoli, M.C. Technological strategies to reduce acrylamide levels in heated foods. *Food Eng. Rev.* **2009**, *1*, 169. [CrossRef]
16. Anese, M.; Suman, M.; Nicoli, M.C. Acrylamide removal from heated foods. *Food Chem.* **2010**, *119*, 791–794. [CrossRef]
17. Taeymans, D.; Wood, J.; Ashby, P.; Blank, I.; Studer, A.; Stadler, R.H.; Gondé, P.; Eijck, P.; Lalljie, S.; Lingnert, H.; et al. A review of acrylamide: An industry perspective on research, analysis, formation and control. *Crit. Rev. Food Sci. Nutr.* **2004**, *44*, 323–347. [CrossRef] [PubMed]
18. Biedermann, M.; Noti, A.; Biedermann-Brem, S.; Mozzetti, V.; Grob, K. Experiments on acrylamide formation and possibilities to decrease the potantial of acrylamide formation in potatoes. *Mitt. Lebensm. Hyg.* **2002**, *93*, 668–687.
19. Elmore, J.S.; Koutsidis, G.; Dodson, A.T. Measurements of acrylamide and its precursors in potato, wheat, and rye model systems. *J. Agric. Food Chem.* **2005**, *53*, 1286–1293. [CrossRef] [PubMed]
20. Hedegaard, R.V.; Frandsen, H.; Granby, K.; Apostolopoulou, A.; Skibsted, L.H. Model studies on acrylamide generation from glucose/asparagine in aqueous glycerol. *J. Agric. Food Chem.* **2007**, *55*, 486–492. [CrossRef] [PubMed]
21. Biedermann, M.; Grob, K. Model studies on acrylamide formation in potato, wheat flour and corn starch; ways to reduce acrylamide contents in bakery ware. *Mitt. Lebensm. Hyg.* **2003**, *94*, 406–422.

22. Bråthen, E.; Knutsen, S.H. Effect of temperature and time on the formation of acylamide in starch-based and cereal model systems, flat breads and bread. *Food Chem.* **2005**, *92*, 693–700. [CrossRef]

23. Surdyk, N.; Rosén, J.; Andersson, R.; Åman, P. Effects of asparagine, fructose and baking conditions on acrylamide content in yeasts-leavened wheat bread. *J. Agric. Food Chem.* **2004**, *52*, 2047–2051. [CrossRef]

24. Bergamini, E.; Catellani, D.; Dall'Asta, C.; Galaverna, G.; Dossena, A.; Marchelli, R.; Suman, M. Fate of Fusarium mycotoxins in the cereal product supply chain: The deoxynivalenol (DON) case within industrial bread-making technology. *Food Addit. Contam.* **2010**, *27*, 677–687. [CrossRef]

25. Generotti, S.; Cirlini, M.; Malachova, A.; Sulyok, M.; Berthiller, F.; Dall'Asta, C.; Suman, M. Deoxyivalenol and deoxynivalenol-3-glucoside mitigation strategies: Effective experimental design within industrial rusk-making technology. *Toxins* **2015**, *7*, 2773–2790. [CrossRef] [PubMed]

26. Suman, M.; Manzitti, A.; Catellani, D. A design of experiments approach to studying deoxynivalenol and deoxynivalenol-3-glucoside evolution throughout industrial production of wholegrain crackers exploiting LC-MS/MS techniques. *World Mycotoxin J.* **2012**, *5*, 241–249. [CrossRef]

27. Generotti, S. Transformation of Regulated and Emerging Mycotoxins upon Food Processing: From Field to Digestion. Ph.D. Thesis, University of Parma, Parma, Italy, January 2016.

28. Telford, J.K. A brief introduction to design of experiments. *John Hopkins APL Tech. Dig.* **2007**, *27*, 224–232.

29. Calbiani, F.; Careri, M.; Elviri, L.; Mangia, A.; Zagnoni, I. Development and single-laboratory validation of a reversed-phase liquid chromatography-electrospray-tandem mass spectrometry method for the identification and determination of acrylamide in foods. *J. AOAC Int.* **2004**, *87*, 107–115. [PubMed]

30. Gökmen, V.; Açar, O.C.; Köksel, H.; Acar, J. Effects of dough formula and baking conditions on acrylamide and hydroxymethylfurfural formation in cookies. *Food Chem.* **2007**, *104*, 1136–1142. [CrossRef]

31. Graf, M.; Amrein, T.M.; Graf, S.; Szalay, R.; Escher, F.; Amadò, R. Reducing the acrylamide content of a semi-finished biscuit on industrial scale. *LWT-Food Sci. Technol.* **2006**, *39*, 724–728. [CrossRef]

32. Vass, M.; Amrein, T.M.; Schönbächler, B.; Escher, F.; Amadò, R. Ways to reduce acrylamide formation in cracker production. *Czech J. Food Sci.* **2004**, *22*, 19–21. [CrossRef]

33. Rydberg, P.; Eriksson, S.; Tareke, E.; Karlsson, P.; Ehrenberg, L.; Törnqvist, M. Investigations of factors that influence the acrylamide content of heated foodstuffs. *J. Agric. Food Chem.* **2003**, *51*, 7012–7018. [CrossRef]

34. Vattem, D.A.; Shetty, K. Acrylamide in foods: A model for mechanism of formation and its reduction. *Innov. Food Sci. Emerg. Technol.* **2003**, *4*, 331–338. [CrossRef]

35. Wedzicha, B.L.; Mottram, D.S.; Elmore, J.S. Kinetic models as a route to control acrylamide formation in food. In *Chemistry and Safety of Acrylamide in Food*; Friedman, M., Mottram, D., Eds.; Springer: New York, NY, USA, 2015.

36. Farah, D.M.H.; Zaibunnisa, A.H.; Misnawi, J. Optimization of cocoa beans roasting process using Response Surface Methodology based on concentration of pyrazine and acrylamide. *Int. Food Res. J.* **2012**, *19*, 1355–1359.

 *toxins*

Article

Determination of Deoxynivalenol Biomarkers in Italian Urine Samples

Barbara De Santis [1,*], **Francesca Debegnach** [1], **Brunella Miano** [2], **Giorgio Moretti** [3], **Elisa Sonego** [1], **Antonio Chiaretti** [4], **Danilo Buonsenso** [4] **and Carlo Brera** [1]

[1] Reparto di Sicurezza Chimica degli Alimenti, Istituto Superiore di Sanità, 00161 Rome, Italy
[2] Laboratorio di Chimica Sperimentale, Istituto Zooprofilattico Sperimentale delle Venezie, 35020 Legnaro (PD), Italy
[3] Istituto Zooprofilattico Sperimentale dell'Umbria e delle Marche, 06126 Perugia, Italy
[4] Dipartimento Scienze della Salute della Donna, del Bambino e di Sanità Pubblica, Fondazione Policlinico Universitario Agostino Gemelli IRCCS, Roma-Italia-Università Cattolica del Sacro Cuore, 00168 Roma, Italia
* Correspondence: barbara.desantis@iss.it; Tel.: +39-06-499-02367

Received: 31 May 2019; Accepted: 18 July 2019; Published: 25 July 2019

Abstract: Deoxynivalenol (DON) is a mycotoxin mainly produced by *Fusarium graminearum* that can contaminate cereals and cereal-based foodstuff. Urinary DON levels can be used as biomarker for exposure assessment purposes. This study assessed urinary DON concentrations in Italian volunteers recruited by age group, namely children, adolescents, adults, and the elderly. In addition, vulnerable groups, namely vegetarians and pregnant women, were included in the study. To determine the urinary DON, its glucuronide and de-epoxydated (DOM-1) forms, an indirect analytical approach was used, measuring free DON and total DON (as sum of free and glucuronides forms), before and after enzymatic treatment, respectively. Morning urine samples were collected on two consecutive days, from six different population groups, namely children, adolescent, adults, elderly, vegetarians and pregnant women. Total DON was measured in the 76% of the collected samples with the maximum incidences in children and adolescent age group. Urine samples from children and adolescent also showed the highest total DON levels, up to 17.0 ng/mg_creat. Pregnant women had the lowest positive samples per category (40% for day 1 and 43% for day 2, respectively), low mean levels of total DON (down to 2.84 ng/mg_creat) and median equal to 0 ng/mg_creat. Estimation of DON dietary intake reveals that 7.5% of the total population exceeds the TDI of 1 µg/kg bw/day set for DON, with children showing 40% of individuals surpassing this value (male, day 2).

Keywords: mycotoxins; deoxynivalenol; children; adolescents; pregnant women; vegetarians; biomonitoring

Key Contribution: This study showed that total DON was measured in 76% of the collected samples, with the maximum incidences in children and adolescent age groups. The outputs of the statistical model associating food variables and urinary DON showed that an increase of total cereal consumption significantly associated with total DON in urine. In particular, the increase of pasta consumption affects the urinary DON content, confirming the relevant role of pasta in the Italian diet.

---

1. Introduction

Mycotoxins are natural food and feed contaminants, mainly produced by filamentous fungi of genera *Aspergillus*, *Penicillium*, *Fusarium* and *Alternaria* [1]. Mycotoxin production is promoted by either environmental and agronomic factors, while strong weather variability is a key factor for fungal infection, fungal colonization, and mycotoxin loads [2,3]. Among mycotoxins, trichothecenes represent the main group of *Fusarium* toxins commonly found in cereal grains. Deoxynivalenol

(DON) is one of the most widely diffused natural-occurring trichothecene, is a sesquiterpenoid polar organic compound, belonging to the type B trichothecenes since it contains carbonyl group in C-8. *Fusarium graminearum* and *culmorum*, two important causing agents of the Fusarium Head Blight (FHB), are the most important producers of DON [4]. Acting as a virulence factor for cereal infection (namely in wheat, maize, barley, oat and rye), DON jeopardizes cereal grain quantity and quality from agriculture and health perspective [3] being persistent in cereal food derived products.

European Regulation (EC) 1881/2006 [5] has set maximum levels of DON in unprocessed cereals and cereal foods products intended for direct human consumption in the range of 1750 to 200 µg/kg. The scaled down levels of the contamination from the milling products and the processed ones reflect a well-known reduction rate of about 30% because of the cleaning and processing [6,7] but it has also to be considered the DON degradation process that occurs during industrial baking [8]. The legislative provisions serve to control the levels in food. However, surveys carried out worldwide [2] and in Italy [9] confirmed DON occurrence in cereal samples like wheat and maize, including processed products, thereby suggesting an exposure assessment issue. The correlation between DON dietary exposure and DON presence in urine has already been reported and confirmed [10,11]; therefore, the biomonitoring of DON and DON metabolites in urine may constitute a valuable indicator of the dietary exposure. On the basis of the available information derived from the toxicity studies, a temporary tolerable daily intake (TDI) of 1 µg/kg body weight (bw)/day was established in 2002 [12]. On the basis of the more recent available scientific data, this TDI has been recently confirmed by the European Food Safety Authority (EFSA) as a group-TDI for the inclusion of 3-AcDON, 15-AcDON and DON-3-glucoside, as DON plant metabolites [4]. In regards to acute risk, the Joint Expert Committee on Food and Additives (JECFA) [13] concluded that DON is a probable factor for acute pathologies in humans and derived an Acute Reference Dose (ARfD) of 8 µg/kg bw. In 2017, an extensive review on the presence of DON and its metabolites in human urine was carried out [14]. Authors showed that it is ubiquitous and highlighted geographical differences. So far, the metabolic profile of DON in animals has already been established. DON is metabolized to 13-deepoxy (DOM-1) in animals and is excreted via the feces and urine [15,16]. In addition to DOM-1, other conjugated products may be present as excreted metabolites, namely DON-3-glucuronide (DON-3GlcA) and DON-15-glucuronide (DON-15GlcA), which seem to be the major metabolites in humans, along with their iso forms and DON-8,15-hemiketal-8-GlcAc [17], sulfate and sulfonate forms [18,19]. More recently, DON-7-glucuronide (DON-7GlcA) has been tentatively identified and introduced as a new metabolite [20–22] in humans.

The analysis of urinary glucuronides is crucial for the study of trichothecene biomarkers, because approximately 80% of DON excreted via urine is conjugated with glucuronic acid [4]. A number of biomonitoring studies of DON in urine is available, however, different analytical method approaches, different limits of quantification and different ways to express measured mycotoxin (total DON or free DON, not specified, creatinine corrected, etc.) may give difficulties in comparison results. When expressed in ng/mg of creatinine (ng/mg$_{creat}$), DON levels in urine ranged from 0.2 up to 903.7 ng/mg$_{creat}$ [23–28], with a worst case reported for a group of pregnant women from Croatia [21].

In the present study, morning urine samples were collected over two consecutive days from 203 Italian volunteers; the participants were recruited by age group, namely children (3–9 years), adolescents (10–17 years), adults (18–64 years), and elderly (65 years or above). Vulnerable groups, namely vegetarians and pregnant women, were also included in the study. Each participant was asked to collect a first-morning urine sample on two consecutive days and to complete a Food Frequency Questionnaire (FFQ) reporting dietary habits over a 1-month period and a Food Diary (FD) with detailed information about food items consumed on the day preceding the collection of urine samples. Associations between food consumption and urinary DON were assessed using ordered logistic regression models.

To determine the urinary total DON expressed as sum of free and glucuronides forms, an indirect analytical approach, measuring free and total DON, before and after enzymatic treatment, respectively, was fully validated and applied to the collected urine samples [29,30].

This paper affords a distinctive data set detailing the DON exposure of Italian population in an urban setting, providing data on levels of DON in human urine samples collected, as analyzed by liquid chromatography coupled with mass spectrometry (LC-MS).

2. Results and Discussion

2.1. Analytical Method

The method used in this experimental study was previously published by Turner et al. [10] and applied with minor modifications. In particular, in order to reduce the amount of methanol needed for the elution of mycotoxins from the immunoaffinity column (IAC), three volumes were tested, 4 mL as recommended in the work of Turner et al., 2 mL and 1.5 mL. The performances were evaluated in terms of the recovery factor ($n = 3$), the obtained results were $102 \pm 5\%$, $99 \pm 5\%$ and $56 \pm 10\%$, for 4, 2 and 1.5 mL, respectively. The selected volume was 2 mL, for which a good recovery was achieved also permitting the reduction of the amount of organic solvent used and of the time needed for drying the sample in the subsequent analytical step. Since the IAC was designed for DON analysis, the IAC cross reactivity for DOM-1 was also tested ($n = 3$) with good results in terms of recovery ($98 \pm 1\%$). In regards to the determination step, a single injection for DON and DOM-1 was performed instead of the separated chromatographic run proposed in the reference paper [10].

2.2. In-House Validation

Before applying the method to the collected samples, an in-house validation was performed on 4 different contamination levels; the results obtained for the investigated performance parameters are reported in Table 1. The limits of detection and quantification (LoD and LoQ) values set during validation fit with the purpose of having an analytical method that is able to detect low amounts of DON and DOM-1 for exposure assessment purposes. Recovery factors are in the ranges 95–109% and 81–93% for DON and DOM-1, respectively, and the precision, evaluated in terms of Relative Standard Deviation of repeatability (RSDr), is always below 10%. Despite the absence of a specific official reference, the obtained values were considered satisfactory under the criteria of Regulation (EC) 401/2006 [31], which apply to any DON analysis on food. The average values estimated for method uncertainty ranged between 8 and 14% for DON and between 15 and 18% for DOM-1, being in both cases well below the 44%, which is the maximum acceptable value when the Horwitz approach is applied [31].

Table 1. Performance characteristics obtained during the in-house validation of the method.

Validation Parameter	DON	DOM-1
LoD; μg/L	0.25	0.25
Level 1 (LoQ); μg/L$_{urine}$	0.50	0.51
Mean value; μg/L$_{urine}$	0.54	0.43
s_r; μg/L$_{urine}$	0.03	0.03
RSD$_r$; %	5.55	6.98
Recovery; %	109	84
Uncertainty; %	14	15
Level 2; μg/L$_{urine}$	2.50	2.53
Mean value; μg/L$_{urine}$	2.63	2.11
s_r; μg/L$_{urine}$	0.10	0.14
RSD$_r$; %	3.80	6.63
Recovery; %	105	85
Uncertainty; %	8	18
Level 3; μg/L$_{urine}$	12.50	12.63
Mean value; μg/L$_{urine}$	12.16	11.65
s_r; μg/L$_{urine}$	0.33	0.37
RSD$_r$; %	2.71	3.18
Recovery; %	95	93
Uncertainty; %	8	17
Level 4; μg/L$_{urine}$	62.50	50.50
Mean value; μg/L$_{urine}$	63.35	40.49
s_r; μg/L$_{urine}$	3.86	0.48
RSD$_r$; %	6.09	1.19
Recovery; %	101	81
Uncertainty; %	9	14

2.3. Description of Study Population

In total, 406 first morning urine samples were collected from 203 volunteers belonging to six population groups, namely children, adolescents, adults, elderly, vegetarians and pregnant women. All samples were kept at −20 °C until analysis. The volunteers enrolled for the study are reported, with their anthropometric data, grouped by category, gender and age in Table 2.

The recruitment of subjects was subdivided between Istituto Superiore di Sanità (ISS) and the hospital "Agostino Gemelli (UCSC)", with ISS responsible of the recruitment of adults, elderly and vegetarians and UCSC of children and adolescents by the clinics of Pediatric Unit and pregnant women from the clinics of Gynecology and Obstetrics outpatients. The selection of different groups of the population was specifically requested by EFSA while being considered as an asset for the study due to having different potential metabolic susceptibility, body weights, dietary habits, and consumption rates. Pediatric biological systems and detoxification processes might be widely different, causing an amplification of children's susceptibility to hazards that would have negative consequences [32]. It should be noted that the cereal-based products consumption intakes figures of children in Italy are comparable with those of adults, showing, in addition, that children are consumers of a higher variety of those food products than adults (data from the FFQ and food diary, not shown). These two aspects provide children with, at least, the same exposure risk to that of adults. As for pregnant women, due to a decreased immunocompetence as a consequence of hormone levels changes and their complex and multifactorial interplay [33], this population group is of special interest for its possible susceptibility to DON, which has a recognized impact on the immune response system. As for vegetarians, they were included to attempt to verify and to measure if a special diet, which could be enriched by cereal based products, is a factor that may affect exposure estimates. However, the literature shows that these subgroups were already enrolled as special populations in other DON biomonitoring surveys. Piekkola et al., and Hepworth et al., [26,34] indicated the potential risk to mothers and their babies from DON exposure during pregnancy. Pestka highlighted the need to study the relationship between DON consumption and possible growth effects in susceptible populations such as children and vegetarians [35].

During the recruitment, UCSC initially encountered difficulties for pregnant women relating to a general concern from potential participants regarding the possible outputs of the study, abandonment of the study after initial recruitment, and incorrect collection of urine samples and compilation of questionnaires from participants. Therefore, ISS supported the UCSC hospital in its interactions with under-recruited subgroups (pregnant women and vegetarians). In regards to vegetarians, it was quite difficult to recruit male volunteers to the study since they constitute a low percentage of the vegetarian population.

The evaluation of the ethic aspects related to the study protocol and its approval by an Ethics Committee was requested and obtained before the starting date of the study. The UCSC ethical approval was granted by the Local Ethics Committee on 23 March 2014 (for the Gynecology Unit) and 7 April 2014 (for the Pediatric Unit). ISS ethical approval was granted by the ISS Ethic Committee on 18 February 2014. Informed consent was provided by the participants during their first visit. Approval code: Prot. PRE 84/14 and Prot. CE 14/413.

Table 2. Anthropometric data and number of individuals recruited in Italy grouped by category, gender and age.

Group Category	Gender [a]	N [b]	Weight, kg	Height, cm	BMI [c], kg/m^2
Children (3–9 years)	F	20	27	124	17.3
	M	20	26	119	17.2
Adolescent (10–17 years)	F	20	52	163	19.4
	M	20	63	173	20.7
Adults (18–64 years)	F	15	63	167	22.4
	M	16	77	179	24.0
Elderly (>65 years)	F	10	63	161	24.4
	M	9	70	170	24.3
Vegetarians	F	15	61	166	21.9
	M	16	75	176	24.9
Pregnant women	F	42	66	164	24.6

[a] F, Female; M: Male; [b] N: number of subjects; [c] BMI, Body Mass Index.

2.4. DON Biomarker Levels in Urine Samples

For each collected urine sample three analyses were performed, one for free and one for total DON and DOM-1, the first and second corresponding to before and after enzymatic treatment, with the third for creatinine. The results obtained are reported in Table 3, namely mean values, median (P50) and interquartile (IQR) values are listed as non-adjusted and creatinine-adjusted total DON. In Table 3 the percentage of samples above LOD for total DON is also reported for each category. Moreover, in Table 3 the percentage contribution of free DON and DON-GlcA to total DON is reported.

The mean DON level in urine for the total population studied was 7.67 and 7.93 ng/mL, but differences arose when the different categories were separately taken into account. Among the selected group, pregnant women had the lowest positive samples per category (40% for day 1 and 43% for day 2, respectively), low mean levels of total DON (4.37 ng/mL and 2.70 ng/mL on day 1 and day 2, respectively), and a median equal to 0 ng/mL. Conversely, urine from children and adolescents showed the highest concentrations of total DON, up to 75.9 ng/mL. The vegetarian group, apart from females on day 1, showed median values very close to the mean values, indicating that the values are close to normal distribution.

The median values of total DON in the morning urine ranged from 0 ng/mL for pregnant women on day 1 and day 2 to 12.60 ng/mL for adults male on day 2. However, when the creatinine adjustment was taken into account the median values ranged from 0 ng/mg$_{creat}$ for pregnant women (day 1 and day 2) to 13.0 ng/mg$_{creat}$ for children male (day 2), while the median value for adults male on day 2 decremented to 6.84 ng/mg$_{creat}$. These results emphasize the importance of creatinine adjustment for the correct interpretation of results.

Regarding the DOM-1 content, only 6 samples out of the 406 analyzed urines (1.5%) had detectable levels of this toxin (ranges <LOD-2.6 ng DOM-1/mg$_{creat}$ and <LOD-1.7 ng DOM-1/mg$_{creat}$, for day 1 and day 2, respectively), confirming that this is a minor route for DON in human metabolism.

In view of the ambiguous statistical differences among males and females' DON levels found in literature [11,30,36–38], DON ng/mg$_{creat}$ content was checked between sexes in the general population excluding pregnant women. No statistical differences were highlighted for any of the age group (Mann-Whitney test). Conversely, a difference statistically significant was obtained comparing women adult group with pregnant women pointing out that pregnant women showed DON ng/mg$_{creat}$ levels lower than adult females in general (*p*-value 0.025).

Comparing the results obtained within the EFSA study [39], the median concentrations of total DON adjusted for creatinine in morning urine in Italy and Norway were quite similar within similar population groups, whereas the median concentrations in the corresponding population groups in the UK were approximately 3-fold higher than in Italy and Norway. The Italian results are also in line with the values published by Solfrizzo et al. [30], which reported for total DON a median of 10.32 ng/mL.

Regarding the contribution of free DON and DON-GlcA to total DON, throughout all the population groups, the DON-GlcA represented 66% and 71% of the total DON for day 1 and day 2 respectively, confirming that the glucuronidation is an important route for DON excretion, as has already been reported in other studies [40].

Table 3. Non-adjusted and creatinine-adjusted total DON concentrations in urine samples by day and sex in sampled population groups, and contribution of free DON and DON-GlcA to the mean total DON concentration.

Age Group	Day	Gender [a]	Samples Above LoQ (%)	Total DON (ng/mL$_{urine}$)					Total DON (ng/mg$_{creat}$)		
				Mean	P50	IQR	Free DON (%)	DON-GlcA (%)	Mean	P50	IQR
Children (3–9 Years)	1	F (20)	90	11.4	5.80	6.91	24	76	13.5	7.31	11.20
	1	M (20)	95	10.3	9.46	13.11	27	73	12.4	9.77	8.16
	2	F (20)	90	8.70	6.49	8.69	29	71	12.9	9.42	11.65
	2	M (20)	95	14.0	11.9	12.60	36	64	17.0	13.0	12.05
Adolescents (10–17 Years)	1	F (20)	80	9.83	5.36	11.89	24	76	9.38	6.20	11.37
	1	M (20)	100	12.9	11.4	12.71	34	66	15.6	9.30	17.54
	2	F (20)	85	9.19	5.12	7.40	15	85	12.6	8.06	9.74
	2	M (20)	90	12.3	11.0	12.17	30	70	11.3	9.96	11.89
Adults (18–64 Years)	1	F (15)	80	5.34	4.18	7.42	44	56	5.33	3.41	6.83
	1	M (16)	88	5.35	5.08	7.12	58	42	4.30	3.76	4.20
	2	F (15)	80	9.01	5.44	7.65	23	77	8.35	5.54	5.87
	2	M (16)	81	11.8	12.6	13.94	45	55	7.60	6.84	11.81
Elderly (>65 Years)	1	F (10)	70	5.07	3.18	6.01	41	59	6.69	5.13	6.10
	1	M (10)	90	4.93	4.40	4.10	41	59	6.37	6.87	8.79
	2	F (10)	60	5.51	4.40	3.25	35	65	7.79	5.16	3.84
	2	M (10)	80	6.16	5.59	3.40	20	80	12.4	8.09	3.84
Vegetarians	1	F (15)	80	10.6	9.74	14.74	19	81	16.1	10.1	30.75
	1	M (15)	73	4.10	3.30	5.22	35	65	4.43	3.23	7.85
	2	F (15)	80	5.53	4.17	5.34	23	77	8.47	8.25	12.73
	2	M (15)	73	5.88	3.03	7.79	30	70	4.28	2.25	7.44
Pregnant Women	1	F (42)	40	4.37	0.00	8.20	28	72	6.30	0.00	6.87
	2	F (42)	43	2.70	0.00	3.10	33	67	2.84	0.00	4.91
Total	1	(203)	76	7.67	4.58	8.63	34	66	9.21	5.05	9.81
	2	(203)	75	7.93	5.29	9.26	29	71	9.03	6.07	10.41

[a] F: Female; M: Male; in parenthesis, number of subjects.

2.5. Regression Analysis Between Food Consumption and DON Level in Urine Samples

The logistic regression model was applied to the dataset to assess the effect of the general variables on total DON concentration adjusted for creatinine. The output of the model refers to a unitary variation of the considered variable. With the aim to express the output of the food variable in a more comprehensive way, an increment of 10 grams of cereal food intake (i.e. total food, food category or item) was considered.

To assess the effect of the general variables, the mean values for the total DON content (ng/mg$_{creat}$) on day 1 and day 2 were calculated for each subject and then categorized in tertiles, resulting in only one dependent variable. In regards to the age groups, the odds ratio (OR) to have a higher level of mean DON adjusted for creatinine in adults compared to children is 0.13 (p = 0.000), confirming the critical scenario for the children age group. The OR to have a higher level of mean concentration of total DON adjusted for creatinine in pregnant women compared to non-pregnant women is 0.198 (p = 0.001), in accordance with the very low DON levels in urine for this category. The gender, BMI, physical activity and vegetarian variables were not significant when added into the model.

Regarding the food variables, the analyses were performed by assessing the effect on total DON adjusted for creatinine on day 1 and day 2 of (i) total cereal food intake; (ii) each food group intake; (iii) single food items. While no statistically significant association between total food intake and DON levels was observed for day 1, the increase of 10 g in the total cereal food intake raised the OR by about 2.6% for day 2 (p = 0.027). The effects of the food groups provided significant results for pasta and pasta-like products, while for day 1 and day 2, an increased intake of 10 g caused the OR to have a higher level of concentration of total DON adjusted for creatinine by about 4% (p = 0.047) and 5.5% (p = 0.052), respectively. Considering the food item variables, the increased intake of durum wheat pasta caused the OR to have a higher level of total DON adjusted for creatinine by about 6% (p = 0.023) and 7.9% (p = 0.008), respectively.

2.6. Estimated DON Daily Intake

Starting from the total DON concentration levels measured in the collected urine samples, the Estimated Dietary Intake (EDI) was calculated using the following formula [30]

$$EDI_{DON} = C \times \frac{V}{bw} \times \frac{100}{E} \tag{1}$$

EDI (ng/kg bw/day);
C = total DON concentration in the analyzed urine samples (ng/mL$_{urine}$);
V = mean 24 h human urine volume (1.0 mL per kg of bw per hour for adults [41]; 2.0 mL per kg of bw per hour for children [42]);
bw = body weight reported in the questionnaire;
E = urinary excretion rate of DON in 24 h, 72.3% [10]

The calculated individual exposure followed a non-normal distribution (Shapiro-Wilk test) hence a non-parametric approach has been used. In Table 4, the mean, median (P50), and 95th percentile (P95) are reported. The obtained EDI values were compared with the Tolerable Daily Intake (TDI) set by EFSA for DON at 1000 ng/kg bw [4], in the last column of Table 4, the number of individuals and the percentage exceeding the TDI are reported.

Table 4. Mean, median (P50) and 95th percentile (P95) of EDI (ng/kg bw/day) calculated for DON are reported based on the category, gender and age group, together with the number and percentage of individuals exceeding the TDI.

Age Group	Day	Gender	Mean	P50	IQR	P95	>TDI
Children (3–9 Years)	1	F (20)	757	385	459	3265	4 (20%)
		M (20)	683	628	870	1975	6 (30%)
	2	F (20)	577	431	577	1765	3 (15%)
		M (20)	937	791	837	2457	8 (40%)
Adolescents (10–17 Years)	1	F (20)	328	178	393	1261	2 (10%)
		M (20)	427	379	421	1359	1 (5%)
	2	F (20)	296	170	246	1019	1 (5%)
		M (20)	409	367	403	1030	1 (5%)
Adults (18–64 Years)	1	F (15)	177	139	247	513	0
		M (16)	170	168	237	446	0
	2	F (15)	299	181	254	1172	1 (7%)
		M (16)	411	416	463	1685	1 (7%)
Elderly (>65 Years)	1	F (10)	104	124	132	284	0
		M (10)	137	141	136	299	0
	2	F (10)	169	146	212	515	0
		M (10)	181	189	129	515	0
Vegetarians	1	F (15)	360	323	472	1555	1 (7%)
		M (15)	186	112	194	561	0
	2	F (15)	181	138	174	549	0
		M (15)	235	119	259	815	0
Pregnant Women	1	F (42)	154	0	271	630	0
	2	F (42)	96	0	103	519	0
Total	1	(203)	329	170	286	1144	14 (7%)
	2	(203)	343	181	310	1128	15 (8%)

The EDI mean values for the total population of the study are quite low, representing around the 30% of the TDI reported by EFSA, and the percentage went down to 20% when the median was considered. The scenario was more differentiated when the single categories were considered. The highest EDI values were obtained for children and to a minor extent for the adolescent age group, with mean values ranging from 577 to 937, and from 296 to 427 ng/kg bw/day for children and adolescent, respectively. Also, the number of individuals exceeding the TDI was higher for this age group, with the highest percentage being the 40% of the male children category on day 2. It is important to note that the two highest estimated EDI's were obtained from female children on day 1 (5036 ng/kg bw/day) and male children on day 2 (3193 ng/kg bw/day), both values were below the ARfD (8000 ng/kg bw/day). As far as chronic exposure is concerned, all the values over the TDI should be duly considered as possible concern for public health as reported by EFSA, especially for vulnerable groups such as infants, toddlers and other children [4].

For a better visualization of the distribution of the obtained results, the calculated individual EDIs are reported in Figure 1. The depicted scenario for children and adolescent must be considered while taking into consideration body weights, homeostasis water balance (i.e., average food and beverage water input, urine feces, skin water output) and food consumption rates, especially for the children, that are not dissimilar when compared to the ones of the rest of the population, producing an unfavorable body weight/intake ratio.

The concentration values obtained in this study were also used by EFSA for exposure estimations. When comparing EFSA exposure estimates from biomarkers for adolescents, adult, elderly with corresponding mean averaged values obtained in this study (data not shown), exposure scenarios are confirmed. Discrepancies arise for children, and in particular, EFSA [4] obtained lower exposure estimates from biomarkers. These differences are due to different urine volumes considered for the

24 h, since EFSA used 0.5 L of urine for children during the 24 h, while in this study, calculations were made depending on the body weight of the subject [42], leading to a urine volume in the 24 h that ranged from 0.71 L/day to 2.49 L/day (an average of 1.25 L/day). The estimation made for the total Italian population in this paper, 329 and 343 ng/kg bw/day for day 1 and day 2 respectively, is also comparable with the probable daily intake (PDI) of 590 ng/kg bw/day reported by Solfrizzo [30], which was calculated considering a 50% excretion rate and 1.5 L for total urine in the 24 h for all of the recruited population.

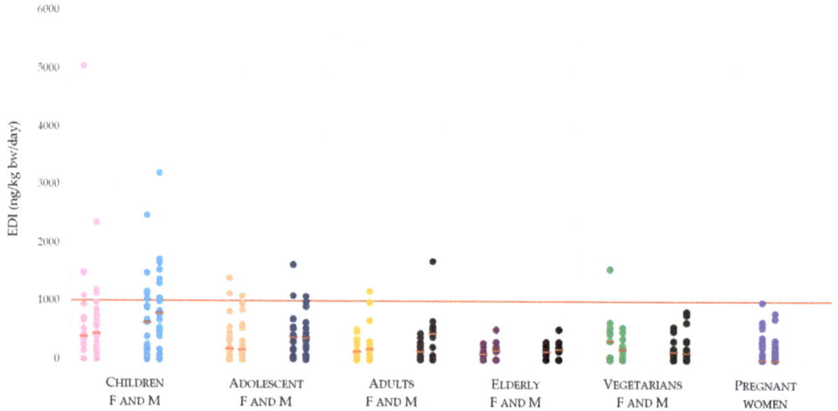

Figure 1. Distribution of the calculated individual EDI around the median value (red small line) and compared to the DON TDI (red line).

3. Conclusions

The high incidence of DON in urine confirms its ubiquitous presence in cereal food products in Italy. However, the results of this study showed moderate mean levels of DON in urine samples of the studied cohort, despite differences that emerged when the different categories were considered, with children being the most susceptible group. The critical scenario depicted for children was also confirmed when the estimated daily intake was considered, showing the highest mean values for this age group (close to TDI and over in the case of P95), and the highest percentage of individuals exceeding the TDI, while for the total population the mean EDI represented around the 30% of the TDI. The higher exposure estimated for children and, to a minor extent for adolescents, can be explained by taking their unfavorable body weight/intake ratios into consideration.

In this study, the contribution of free DON and DON-GlcA to total urinary DON was also investigated and the results confirmed that DON-GlcA is the major DON metabolites for humans, on the other hand the very limited number of samples with DOM-1 above LOD confirms that this is a minor route for DON in the human metabolism.

Statistical analysis confirmed that age significantly affected urinary DON concentration (the difference between children and adults). Considering the outputs of the statistical model for food variables, increasing total cereal consumption was significantly associated with total DON in urine and in particular the increase of pasta consumption affected the urinary DON content more than the increased intake of other studied food items, confirming the relevant role of pasta in the Italian diet.

The results obtained in this study underpin the need for other studies in order to collect data for providing a more comprehensive exposure assessment of the Italian population. Moreover, support for the biomarker approach in exposure assessment is represented by the availability of either validated analytical methods and harmonized references for those critical figures (such as urine outputs or excretion rates), which sensibly influence the exposure scenarios.

4. Materials and Methods

4.1. Analytical Method

4.1.1. Chemicals and Reagents

Methanol HPLC grade (Carlo Erba Reagenti, Cornaredo, MI, Italy) and ultrapure water (Millipore, Burlington, MA, USA) were used for the LC-MS analysis.

In order to perform the planned validation study and the analytical work on collected urine samples, the standard of DON (product number: D0156, 1mg), U-[$^{13}C_{15}$]-DON (99.0% ^{13}C) and DOM-1 (product number: 34135, 2 mL) were purchased from Sigma (Saint Louis, MI, USA). The enzyme β-glucuronidase (Type IX-A from *E. coli*; product number: G7396—2MU) was purchased from Sigma too. The IAC DONtest WBTM (product number: G1066) were purchased from Vicam (Milford, MA, USA).

4.1.2. Sample Preparation

Urinary DON and metabolites concentrations were measured using a two-step process. Stored urine samples were allowed to thaw, and were then centrifuged (1790 g, +4 °C, 15 min). For each participant, two aliquots (1 mL) were prepared by mixing ^{13}C-DON internal standard solution, to provide a final concentration of 20 ng/mL.

Aliquot 1 was used to determine total DON concentrations, defined as the sum of glucuronide metabolites and free DON. To measure DON-glucuronides and free DON, each sample was adjusted to pH 6.8 with drop wise addition of KOH or HCl and digested using β-glucuronidase solution (23,000 U/mL, 250 µL) in a shaking water bath at 37 °C for 18 h, ensuring a gentle mixing. After this pre-treatment, the samples were diluted to a final 4 mL with phosphate buffered saline (PBS, pH 7.4). The diluted urine sample was passed through a wide bore DON IAC using a Visiprep^{TM} vacuum manifold (Sigma-Aldrich, Saint Louis, MI, USA). DON was eluted from columns with methanol (2 mL) and extracts were dried at 40 °C under a gentle stream of nitrogen and reconstituted in 90% methanol (250 µL) before LC-MS analysis. DOM-1 was quantified on the same aliquot and analyzed for DON-GlcA. Aliquot 2 was used to assess free DON using the aforementioned procedure, but without any β-glucuronidase treatment.

Urine creatinine was assessed by the enzymatic method described by Mazzachi et al. [43].

4.1.3. LC-MS Determination

Chromatographic separation of DON and DOM-1 was performed using UHPLC (Ultra-High-Performance Liquid Chromatography) with Waters RP Acquity BEH C18 column (100 × 2.1 mm, 1.7 µm, Milford, MA, USA) kept at 30 °C, and a mobile phase sequence of 10 minutes duration, starting with 20% MeOH, changing to wash of 75% MeOH after 4.50 min and reverting to 20% MeOH after 8 min (flow rate 0.350 mL/min with a volume injection of 10 µL). DON elutes at 2.5 minutes under these conditions, while DOM-1 elutes at 4.5 min.

The mass spectrometric analysis was carried out with a Quattro Premier XE (Waters Milford, MA, USA) in SIR (Selective Ion Recording) acquisition mode with an ESI (ElectroSpray Ionization) interface. The analysis was performed in positive ion-mode. The following mass spectrometer conditions were optimized by direct infusion of DON and DOM-1 standard solutions: capillary voltage 4.0 kV, desolvation gas flow rate 500 L/h at 350 °C, source temperature 110 °C, cone voltage 40 V. The monitored masses for DON were 319.2 and 334.2 m/z corresponding to [DON-Na]$^+$ and [^{13}C-DON-Na]$^+$ respectively, and 303.2 m/z for [DOM-1-Na]$^+$.

For quantification purposes, calibration curves with a labeled internal standard for DON, and an external standard approach for DOM-1 were used. The curves covered the range 2–250 ng/mL, corresponding to 0.50–62.5 µg/L$_{urine}$ for DON, and 2–200 ng/mL corresponding to 0.50–50 µg/L$_{urine}$ for DOM-1. An acceptability criterion of $R^2 > 0.995$ was applied during routine analysis.

DON-GlcA values were estimated indirectly by subtracting free DON values from total DON values for each analyzed urine sample.

4.2. In-House Validation

The in-house validation was performed in accordance with the Eurachem guideline [44]. Selectivity and specificity were guaranteed by the clean-up step with the IAC containing specific antibodies for the selected mycotoxins. LoD and LoQ were identified by the injection of diluted standard solutions, the requirements were a S/N = 3 for LOD and a S/N = 10 for LoQ. The LoQ was included in the validated contamination levels. The validation was performed on 4 different contamination levels for DON and DOM-1 by repeated analyses on spiked urine samples. Trueness was evaluated in terms of recovery factors, while precision was estimated in terms of Relative Standard Deviation of repeatability (RSD_r) calculated on repeated analyses for each contamination level. Expanded uncertainty was also estimated by a metrological approach in accordance with Eurachem guideline [45]. The combined standard uncertainty was calculated by a metrological approach by summing up the standard uncertainty contributions from repeatability (Type A), recovery (Type A), pipetting volumes (Type B), and calibration (Type A). By applying a coverage factor of 2, the expanded uncertainty accounted for the 95% confidence interval.

4.3. Study Design

The dataset in the present analysis represents a subset of data collected for a larger study entitled "Experimental study of deoxynivalenol biomarkers in urine" conducted for the European Food Safety Authority (EFSA) GP/EFSA/CONTAM/2013/04 [39]. In brief, this study explored the occurrence of DON and its metabolites in urine from different population groups (children, adolescents, adults, elderly, and pregnant women; total n = 635) in three European countries (UK, Italy, and Norway) and the relationships between urinary DON levels and its metabolites and self-reported dietary intake of cereal-based food items.

Recruitment of Participants and Urine Sample Collection

A target sample size of at least 200 individuals was established. The population groups included in the study were divided according to the age groups used within the EFSA Comprehensive European Food Consumption Database [46]. A relatively higher number of potentially more susceptible population groups such as children, adolescents, vegetarian and pregnant women was planned to be included. The planned sampled population included six different subgroups for recruitment, i.e., children (aged 3–9, 20%), adolescents (aged 10–17, 20%), adults (aged 18–64, 10%), elderly (aged above 65, 15%), vegetarians (15%) and pregnant women (20%).

Exclusion criteria included subjects not being able to give informed consent or complete the questionnaire, individuals affected by acute pathologies and any chronic illness (chronic renal, hepatic or cardiac problems, cancer), with chronic gastrointestinal conditions (e.g., celiac disease), gluten sensitivity or eating disorders, such as food allergies and those subjects recently on a weight loss diet, depression and psychosis, or hospitalized subjects within 3 months of admission. Inclusion criteria at the specialized recruitment centers (hospitals, clinics, institutions) required only healthy people, either not being on any medication or stable medication (for more than three months) that did not affect appetite (such as oral steroid use). As far as vegetarians, only people following the diet for >1 year and above the age of 18 years were recruited.

Collection equipment and instructions on how to collect and store a first morning urine sample were provided to participants. On two separate days, prior to urine sample collection, participants were required to complete a food diary, reporting food items consumed throughout those days. Participants provided a first morning urine sample the following morning. The urine samples (kept frozen at home) were returned to the clinical trial units after the second day of collection.

4.4. Regression Analysis Between Food Consumption and DON Levels in Urine Samples

4.4.1. Food Frequency Questionnaire and Food Diary

Since a rich cereal-based diet plays a key role in the DON exposure, a semi-quantitative Food Frequency Questionnaire (FFQ) was designed. Type, frequency and quantity of the food consumed were obtained by this FFQ which has been prepared based on a validated questionnaire used in a Spanish study targeted to pregnant women [47]. The adopted FFQ included information about portion size and usual food frequency intake, with a recall period up to one month. In order to make the compilation of the questionnaire easier, photographic examples of portion sizes were also provided to participants. The food list in the adopted FFQ included specific food categories containing all food sources susceptible to DON contamination such as wheat, maize and barley products with emphasis on breads (whole meal, white, soft grain, other), breakfast cereals (high-fiber and other), pasta, pizza, fruit pies, biscuits, buns/cakes and beer. The reported food items reflected the specific Italian food habits. Therefore, in order to get information on the dietary habits of each population group, the FFQ was designed for gathering information on the amount and type of cereals and cereal-based products commonly more frequently consumed by the volunteers in the month prior the study and to capture and verify the most common food sources of DON in the diet. In order to capture such food categories, the EFSA's Food Classification System FoodEx2 database was used [46].

Beside the FFQ, a Food Diary (FD) was also prepared with the aim of collecting information on the intake of the same food items as those included in the FFQ, but referred to the two days immediately before the urine sampling.

A database collecting all the data related to the enrolled subjects concerning individual information (age, gender, BMI, and other information) and food intake was prepared.

4.4.2. Statistical Analysis

Statistical analyses were carried out using STATA/SE 12.0. For descriptive statistics measures (e.g. mean, median) a substitution method was applied. Values below LOD were substituted with 0. Shapiro-Wilk test was used for testing of normality distribution. Two sample Wilcoxon rank sum test (Mann-Whitney test) was used to compare within categories (e.g., Male/Female or Vegetarian Y/N).

To assess the effect of all the observed variables on the total DON content (expressed as ng/mg$_{creat}$) an ordered logistic regression model was used. As variables, the model used a selection of the information gathered from the questionnaires including general variables such as age, gender and BMI, and food variables, namely the different food items reported on the FFQ and the FD. The model was run matching independent variables with the dependent variable. The variables that were shown to be significant were matched together to assess how much of the observed phenomenon is explained by the variables considered. When variables were significant, also interactions were considered. In order to apply the ordered logistic regression model, a categorization in tertiles (33th, 66th, 100th) of the dependent variable, namely the total DON content (ng/mg$_{creat}$) on day 1 and day 2 was performed.

The odds ratio (OR) is used to quantify the extent of the association of the examined variable on the total DON (ng/mg$_{creat}$) values.

A variable is considered significant when the null hypothesis of the relative coefficient of the variable is null with a significance level set to 0.95 (1-alpha). For this purpose, the *p* value was used. When the *p* value is higher than 0.05, the examined variable is considered to not be significant, conversely if the *p* value is lower than 0.05 the examined variable is considered to be significant.

Author Contributions: Conceptualization, B.D.S. and C.B.; Data curation, B.D.S., F.D. and G.M.; Formal analysis, F.D., B.M., responsible for the biological sampling (urine samples collection), A.C. and D.B.; Validation, B.D.S., F.D., B.M. and E.S.; Writing—original draft, B.D.S., F.D. and E.S.; Writing—review & editing, C.B.

Funding: The data presented in this study is extracted from a larger study funded by the European Food Safety Authority (EFSA) call GP/EFSA/CONTAM/2013/04 into the "Experimental Study of Deoxynivalenol Biomarkers in

Urine". Sole responsibility lies with the author and the Authority is not responsible for any use that may be made of the information contained therein.

Acknowledgments: Marco Finocchietti, Gabriele Moracci, Maria Cristina Barea Toscan and Giuliana Verrone are acknowledged for their technical assistance.

Conflicts of Interest: The authors declare no conflict of interest.

References

1.   Puntscher, H.; Kütt, M.L.; Skrinjar, P.; Mikula, H.; Podlech, J.; Fröhlich, J.; Marko, D.; Warth, B. Tracking emerging mycotoxins in food: Development of an LC-MS/MS method for free and modified Alternaria toxins. *Anal. Bioanal. Chem.* **2018**, *410*, 4481–4494. [CrossRef] [PubMed]
2.   Ferrigo, D.; Raiola, A.; Causin, R. Fusarium Toxins in Cereals: Occurrence, Legislation, Factors Promoting the Appearance and Their Management. *Molecules* **2016**, *21*, 627. [CrossRef] [PubMed]
3.   Audenaert, K.; Vanheule, A.; Höfte, M.; Haesaert, G. Deoxynivalenol: A Major Player in the Multifaceted Response of Fusarium to Its Environment. *Toxins* **2014**, *6*, 1–19. [CrossRef] [PubMed]
4.   European Food Safety Authority (EFSA). Risks to human and animal health related to the presence of deoxynivalenol and its acetylated and modified forms in food and feed. *EFSA J.* **2017**, *15*, 4718–5093. [CrossRef]
5.   Commission Regulation (EC). No 1881/2006 setting maximum levels for certain contaminants in foodstuffs. *Off. J. Eur. Union* **2006**, *364*, 5–24.
6.   Visconti, A.; Haidukowski, M.; Pascale, M.; Silvestri, M. Reduction of Deoxynivalenol during durum wheat processing and spaghetti cooking. *Toxicol. Lett.* **2004**, *153*, 181–189. [CrossRef] [PubMed]
7.   Brera, C.; Peduto, A.; Debegnach, F.; Pannunzi, E.; Prantera, E.; Gregori, E.; De Giacomo, M.; De Santis, B. Study of the influence of the milling process on the distribution of Deoxynivalenol content from the caryopsis to cooked pasta. *Food Control* **2012**, *32*, 309–312. [CrossRef]
8.   Stadler, D.; Lambertini, F.; Woelflingseder, L.; Schwartz-Zimmermann, H.; Marko, D.; Suman, M. The Influence of Processing Parameters on the Mitigation of Deoxynivalenol during Industrial Baking. *Toxins* **2019**, *11*, 317. [CrossRef]
9.   Amoriello, T.; Belocchi, A.; Quaranta, F.; Ripa, C.; Melini, F.; Aureli, G. Behaviour of durum wheat cultivars towards deoxynivalenol content: A multi-year assay in Italy. *Ital. J. Agron.* **2017**, *13*, 12–20. [CrossRef]
10.   Turner, P.C.; Hopton, R.P.; Lecluse, Y.; White, K.L.M.; Fisher, J.; Lebailly, P. Determinants of urinary deoxynivalenol and de-epoxy deoxynivalenol in male farmers from Normandy, France. *J. Agric. Food Chem.* **2010**, *58*, 5206–5212. [CrossRef]
11.   Vidal, A.; Claeys, L.; Mengelers, M.; Vanhoorne, V.; Vervaet, C.; Huybrechts, B.; De Saeger, S.; De Boevre, M. Humans significantly metabolize and excrete the mycotoxin deoxynivalenol and its modified form deoxynivalenol-3-glucoside within 24 hours. *Sci. Rep.* **2018**, *27*, 5255–5266. [CrossRef]
12.   Scientific Committee on Food (SCF). Opinion of the Scientific Committee on Food on Fusarium toxins. Part 6: Group Evaluation of T-2 Toxin, HT-2 Toxin, Nivalenol and Deoxynivalenol. 2002. Available online: https://ec.europa.eu/food/sites/food/files/safety/docs/sci-com_scf_out123_en.pdf (accessed on 30 May 2019).
13.   Joint Expert Committee on Food and Additives (JECFA). Evaluation of certain food additives and contaminants. In *Report of the Seventy-Second Meeting of the Joint FAO/WHO Expert Committee on Food Additives*; WHO: Geneva, Switzerland, 2010.
14.   Chen, L.; Yu, M.; Wu, Q.; Peng, Z.; Wang, D.; Kuca, K.; Yao, P.; Yan, H.; Nussler, A.K.; Liu, L.; et al. Gender and geographical variability in the exposure pattern and metabolism of deoxynivalenol in humans: A review. *J. Appl. Toxicol.* **2017**, *37*, 60–70. [CrossRef]
15.   Prelusky, D.B.; Hamilton, R.M.; Trenholm, H.L.; Miller, J.D. Tissue distribution and excretion of radioactivity following administration of 14C-labeled deoxynivalenol to White Leghorn hens. *Fundam. Appl. Toxicol.* **1986**, *7*, 635–645. [CrossRef]
16.   Prelusky, D.B.; Trenholm, H.L. Tissue distribution of deoxynivalenol in swine dosed intravenously. *J. Sci. Food Agric.* **1991**, *39*, 748–751. [CrossRef]

17. Schwartz-Zimmermann, H.E.; Hametner, C.; Nagl, V.; Fiby, I.; Macheiner, L.; Winkler, J.; Dänicke, S.; Clark, E.; Pestka, J.J.; Berthiller, F. Glucuronidation of deoxynivalenol (DON) by different animal species: Identifcation of iso-DON glucuronides and iso-deepoxy-DON glucuronides as novel DON metabolites in pigs, rats, mice, and cows. *Arch. Toxicol.* **2017**, *91*, 3857–3872. [CrossRef]

18. Warth, B.; Del Favero, G.; Wiesenberger, G.; Puntscher, H.; Woelflingseder, L.; Fruhmann, P.; Sarkanj, B.; Krska, R.; Schuhmacher, R.; Adam, G.; et al. Identification of a novel human deoxynivalenol metabolite enhancing proliferation of intestinal and urinary bladder cells. *Sci. Rep.* **2016**, *6*, 33854. [CrossRef]

19. Schwartz-Zimmermann, H.E.; Hametner, C.; Nagl, V.; Slavik, V.; Moll, W.-D.; Berthiller, F. Deoxynivalenol (DON) sulfonates as major DON metabolites in rats: From identification to biomarker method development, validation and application. *Anal. Bioanal. Chem.* **2014**, *406*, 7911–7924. [CrossRef]

20. Maul, R.; Warth, B.; Kant, J.S.; Schebb, N.H.; Krska, R.; Koch, M.; Sulyok, M. Investigation of the hepatic glucuronidation pattern of the Fusarium mycotoxin deoxynivalenol in various species. *Chem. Res. Toxicol.* **2012**, *25*, 2715–2717. [CrossRef]

21. Sarkanj, B.; Warth, B.; Uhlig, S.; Abia, W.A.; Sulyok, M.; Klapec, T.; Krska, R.; Banjari, I. Urinary analysis reveals high deoxynivalenol exposure in pregnant women from Croatia. *Food Chem. Toxicol.* **2013**, *62*, 231–237. [CrossRef]

22. Warth, B.; Sulyok, M.; Berthiller, F.; Schuhmacher, R.; Krska, R. New insights into the human metabolism of the Fusarium mycotoxins deoxynivalenol and zearalenone. *Toxicol. Lett.* **2013**, *220*, 88–94. [CrossRef]

23. Ali, N.; Blaszkewicz, M.; Degen, G.H. Assessment of deoxynivalenol exposure among Bangladeshi and German adults by a biomarker-based approach. *Toxicol. Lett.* **2016**, *258*, 20–28. [CrossRef]

24. Ali, N.; Blaszkewicz, M.; Al Nahid, A.; Rahman, M.; Degen, G.H. Deoxynivalenol Exposure Assessment for Pregnant Women in Bangladesh. *Toxins* **2015**, *7*, 3845–3857. [CrossRef]

25. Gratz, S.W.; Richardson, A.J.; Duncan, G.; Holtrop, G. Annual variation of dietary deoxynivalenol exposure during years of different Fusarium prevalence: A pilot biomonitoring study. *Food Addit. Contam. Part A Chem. Anal. Control Expo. Risk Assess.* **2014**, *31*, 1579–1585. [CrossRef]

26. Hepworth, S.J.; Hardie, L.J.; Fraser, L.K.; Burley, V.J.; Mijal, R.S.; Wild, C.P.; Azad, R.; McKinney, P.A.; Turner, P.C. Deoxynivalenol exposure assessment in a cohort of pregnant women from Bradford, UK. *Food Addit. Contam. Part A Chem. Anal. Control Expo. Risk Assess.* **2012**, *29*, 269–276. [CrossRef]

27. Turner, P.C.; Ji, B.T.; Shu, X.O.; Zheng, W.; Chow, W.H.; Gao, Y.T.; Hardie, L.J. A biomarker survey of urinary deoxynivalenol in China: The Shanghai Women's Health Study. *Food Addit. Contam. Part A Chem. Anal. Control Expo. Risk Assess.* **2011**, *28*, 1220–1223. [CrossRef]

28. Wallin, S.; Gambacorta, L.; Kotova, N.; Lemming, E.W.; Nalsen, C.; Solfrizzo, M.; Olsen, M. Biomonitoring of concurrent mycotoxin exposure among adults in Sweden through urinary multi-biomarker analysis. *Food Chem. Toxicol.* **2015**, *83*, 133–139. [CrossRef]

29. Turner, P.C.; Burley, V.J.; Rothwell, J.A.; White, K.L.M.; Cade, J.E.; Wild, C.P. Deoxynivalenol: Rationale for the development and application of urinary biomarker. *Food Addit. Contam. Part A Chem. Anal. Control Expo. Risk Assess.* **2008**, *25*, 864–871. [CrossRef]

30. Solfrizzo, M.; Gambacorta, L.; Visconti, A. Assessment of multi-mycotoxin exposure in southern Italy by urinary multi-biomarker determination. *Toxins* **2014**, *6*, 523–538. [CrossRef]

31. Commission Regulation (EC). No 401/2006 of 23 February 2006 laying down the methods of sampling and analysis for the official control of the levels of mycotoxins in foodstuffs. *Off. J. Eur. Union* **2006**, *70*, 12–34.

32. Hines, R.N.; Sargent, D.; Autrup, H.; Birnbaum, L.S.; Brent, R.L.; Doerrer, N.G.; Cohen Hubal, E.A.; Juberg, D.R.; Laurent, C.; Luebke, R.; et al. Approaches for Assessing Risks to Sensitive Populations: Lessons Learned from Evaluating Risks in the Pediatric Population. *Toxicol. Sci.* **2010**, *113*, 4–26. [CrossRef]

33. Kourtis, A.P.; Read, J.S.; Jamieson, D.J. Pregnancy and infection. *N. Engl. J. Med.* **2014**, *370*, 2211–2218. [CrossRef]

34. Piekkola, S.; Turner, P.C.; Abdel-Hamid, M.; Ezzat, S.; El-Daly, M.; El-Kafrawy, S.; Savchenko, E.; Poussa, T.; Woo, J.C.S.; Mykkänen, H.; et al. Characterisation of aflatoxin and deoxynivalenol exposure among pregnant Egyptian women. *Food Addit. Contam. Part A Chem. Anal. Control Expo. Risk Assess.* **2012**, *29*, 962–971. [CrossRef]

35. Pestka, J.J. Deoxynivalenol: Mechanisms of action, human exposure, and toxicological relevance. *Arch. Toxicol.* **2010**, *84*, 663–679. [CrossRef]

36. Turner, P.C.; White, K.L.; Burley, V.J.; Hopton, R.P.; Rajendram, A.; Fisher, J.; Cade, J.E.; Wild, C.P. A comparison of deoxynivalenol intake and urinary deoxynivalenol in UK adults. *Biomarkers* **2010**, *15*, 553–562. [CrossRef]

37. Rodriguez-Carrasco, Y.; Molto, J.C.; Manes, J.; Berrada, H. Exposure assessment approach through mycotoxin/creatinine ratio evaluation in urine by GC-MS/MS. *Food Chem. Toxicol.* **2014**, *72*, 69–75. [CrossRef]

38. Ediage, E.N.; Di Mavungu, J.D.; Song, S.Q.; Sioen, I.; De Saeger, S. Multimycotoxin analysis in urines to assess infant exposure: A case study in Cameroon. *Environ. Int.* **2013**, *57–58*, 50–59. [CrossRef]

39. Brera, C.; De Santis, B.; Debegnach, F.; Miano, B.; Moretti, G.; Lanzone, A.; Del Sordo, G.; Buonsenso, D.; Chiaretti, A.; Hardie, L.; et al. Experimental study of deoxynivalenol biomarkers in urine. *EFSA Supporting Publ.* **2015**, *12*, 818E. [CrossRef]

40. Turner, P.C.; Hopton, R.P.; White, K.L.M.; Fisher, J.; Cade, J.E.; Wild, C.P. Assessment of deoxynivalenol metabolite profiles in UK adults. *Food Chem. Toxicol.* **2011**, *49*, 132–135. [CrossRef]

41. Klingensmith, M.E.; Aziz, A.; Bharat, A.; Fox, A. *The Washington Manual of Surgery*, 5th ed.; Lippincott Williams & Wilkins: Philadelphia, PA, USA, 2008.

42. Kliegman, R.M.; Geme, J.S. *Nelson: Textbook of Pediatrics*, 20th ed.; Elsevier: Philadelphia, PA, USA, 2015.

43. Mazzachi, B.C.; Peake, M.J.; Ehrhardt, V. Reference range and method comparison studies for enzymatic and Jaffé creatinine assays in plasma and serum and early morning urine. *Clin. Lab.* **2000**, *46*, 53–55.

44. Magnusson, B.; Örnemark, U. (Eds.) *Eurachem Guide: The Fitness for Purpose of Analytical Methods—A Laboratory Guide to Method Validation and Related Topics*, 2nd ed.; Eurachem: Torino, Italy, 2014; ISBN 978-91-87461-59-0.

45. Ellison, S.L.R.; Williams, A. *Eurachem/CITAC Guide: Quantifying Uncertainty in Analytical Measurement*, 3rd ed.; Eurachem: Torino, Italy, 2012.

46. European Food Safety Authority (EFSA). Use of the EFSA Comprehensive European Food Consumption Database in Exposure Assessment. *EFSA J.* **2011**, *9*, 2097–2131. [CrossRef]

47. Vioque, J.; Navarrete-Muñoz, E.M.; Gimenez-Monzó, D.; García-de-la-Hera, M.; Granado, F.; Young, I.S.; Ramón, R.; Ballester, F.; Murcia, M.; Rebagliato, M.; et al. Reproducibility and validity of a food frequency questionnaire among pregnant women in a Mediterranean area. *Nutr. J.* **2013**, *19*, 12–26. [CrossRef]

Article

Fluorescence Polarization Immunoassay for the Determination of T-2 and HT-2 Toxins and Their Glucosides in Wheat

Vincenzo Lippolis [1,*], Anna C. R. Porricelli [1], Erminia Mancini [1], Biancamaria Ciasca [1], Veronica M. T. Lattanzio [1], Annalisa De Girolamo [1], Chris M. Maragos [2], Susan McCormick [2], Peiwu Li [3], Antonio F. Logrieco [1] and Michelangelo Pascale [1]

[1] Institute of Sciences of Food Production (ISPA), National Research Council of Italy (CNR), 70126 Bari, Italy
[2] Mycotoxin Prevention and Applied Microbiology Research Unit, Agricultural Research Service, U.S. Department of Agriculture, Peoria, IL 61604, USA
[3] Key Lab for Mycotoxins Detection, Oil Crops Research Institute of the Chinese Academy of Agricultural Sciences, Wuhan 430062, China
* Correspondence: vincenzo.lippolis@ispa.cnr.it

Received: 30 April 2019; Accepted: 19 June 2019; Published: 1 July 2019

Abstract: T-2 and HT-2 toxins and their main modified forms (T-2 glucoside and HT-2 glucoside) may co-occur in cereals and cereal-based products. A fluorescence polarization immunoassay (FPIA) was developed for the simultaneous determination of T-2 toxin, HT-2 toxin and relevant glucosides, expressed as sum. The developed FPIA, using a HT-2-specific antibody, showed high sensitivity (IC_{50} = 2.0 ng/mL) and high cross-reactivity (100% for T-2 toxin and 80% for T-2 and HT-2 glucosides). The FPIA has been used to develop two rapid and easy-to-use methods using two different extraction protocols, based on the use of organic (methanol/water, 90:10, *v/v*) and non-organic (water) solvents, for the determination of these toxins in wheat. The two proposed methods showed analytical performances in terms of sensitivity (LOD 10 µg/kg) recovery (92–97%) and precision (relative standard deviations ≤13%), fulfilling the criteria for acceptability of an analytical method for the quantitative determination of T-2 and HT-2 toxins established by the European Union. Furthermore, the methods were then validated in accordance with the harmonized guidelines for the validation of screening methods included in the Regulation (EU) No. 519/2014. The satisfactory analytical performances, in terms of intermediate precision (≤25%), cut-off level (80 and 96 µg/kg for the two methods) and rate of false positives (<0.1%) confirmed the applicability of the proposed methods as screening method for assessing the content of these toxins in wheat at the EU indicative levels reported for T-2 and HT-2 toxins.

Keywords: fluorescence polarization immunoassay; T-2 toxin; HT-2 toxin; T-2 glucoside; HT-2 glucoside; wheat; validation study; screening method

Key Contribution: This manuscript reports the first fluorescence polarization immunoassay for the determination of T-2 and HT-2 toxins and relevant glucosides in wheat. Two alternative extraction protocols were proposed and methods were validated in accordance with the Commission Regulation (EU) No. 519/2014.

---

1. Introduction

T-2 toxin (T-2) and its deacetylated form HT-2 toxin (HT-2) are epoxy sesquiterpenoids, classified as type-A trichothecene mycotoxins, produced by several *Fusarium* species, mainly *F. langsethiae* and *F. sporotrichioides*, in cereal grains under cool and moist conditions in the field and after harvesting.

Several studies report the incidence of T-2 and HT-2 mainly in oats, but also in other grains including barley, wheat, maize, rice and soybean, as well as in cereal-based products [1–4]. Being potent inhibitors of DNA, RNA and protein synthesis, T-2 and HT-2 can cause several adverse effects in both humans and animals [3,5]. Due to their toxicity and co-occurrence, the Panel on Contaminants in the Food Chain (CONTAM Panel) of the European Food Safety Authority (EFSA) established a group tolerable daily intake (TDI) of 100 ng/kg body weight per day for the sum of T-2 and HT-2 [3]. Although maximum permitted levels were not established for T-2 and HT-2, the European Union provided indicative levels for the sum of these toxins in cereals and derived products ranging from 15 to 2000 µg/kg [6] from which investigations should be performed to assess changes and trends in human and animal exposure. For unprocessed wheat samples, the indicative level for T-2 and HT-2 was 100 µg/kg (expressed as sum).

Several modified forms of T-2 and HT-2 generated by fungi, plants and mammals were isolated and characterized, as reported in the EFSA Scientific Opinion [7]. Among the possible modifications for these toxins, conjugation with sugars, mainly with glucose, as glucopyranosides play an important role leading to the formation of T-2 glucoside (T-2G) and HT-2 glucoside (HT-2G). Figure 1 reports the chemical structures of T-2, HT-2, T2-G and HT-2G. The presence of T-2G and HT-2G in naturally contaminated cereals, including wheat, oat, maize and barley, have been reported by several authors [8–11] and recently reviewed [12]. Only limited toxicological data are available for T-2G and HT-2G, but they could also be toxic by releasing their aglycones, either during food processing or in the gastrointestinal tract after ingestion [7]. Although the CONTAM Panel found it appropriate to establish a group TDI for T-2 and HT-2 and their modified forms, they concluded that this assumption would determine an overestimation of risk for these toxins; therefore, based on new toxicity studies, the group TDI, only for T-2 and HT-2, was recently amended to 20 ng/kg body weight per day [7]. Moreover, an estimation of human and animal dietary exposure to T-2 and HT-2 has been recently carried out by EFSA. However, due to lack of data, the potential presence of the T-2/HT-2 modified forms was not considered, which could determine an underestimation of the real exposure [4]. For these reasons, the collection of analytical data on T-2 and HT-2, including modified forms, in food and feed commodities is highly needed using analytical methods with an appropriate sensitivity and accuracy [4,11]. Moreover, the availability of analytical methods able to detect simultaneously mycotoxins and their modified forms, even though expressed as the sum, is very useful especially in view of possible future requirements of European Regulations.

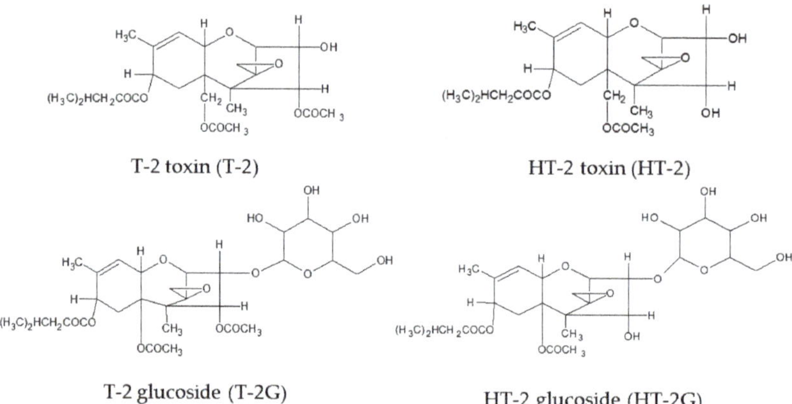

Figure 1. Chemical structures of T-2 and HT-2 toxins and their main modified forms T-2 and HT-2 glucosides.

A large number of analytical methods are available for the determination of T-2 and HT-2 in cereals and derived products, based on high- or ultrahigh-performance liquid chromatography (HPLC and UHPLC) and gas chromatography (GC) [13–15]. Although these methods have high accuracy and sensitivity permitting the determination of T-2 and HT-2 at the indicative levels recommended by the European Commission [6], they cannot be used for the determination of relevant glycosylated forms. For this reason, LC-MS analysis is the most widely used approach for the simultaneous determination of T-2/HT-2 and relevant glucosides, which are also often detected together with other co-occurring mycotoxins, as well as their modified forms [11,16]. However, these analytical methods are expensive, time-consuming and require skilled personnel. For this reason, easy-to-use, rapid, cheap, high-throughput, robust and reliable analytical methods for the simultaneous monitoring of T-2 and HT-2 and relevant glucosides are in great demand for collecting more data and correctly evaluating the real exposure to these toxins. Moreover, in the development of rapid test kits for the detection of mycotoxins the use of non-organic solvents is especially encouraged to provide kits for users who are not familiar with the management and disposal of the organic solvents [17].

In the last decade, several rapid methods have been developed for the detection of T-2 and HT-2 in cereals and derived products, mainly immunochemical methods [13], such as enzyme-linked immunosorbent assays [18–21], lateral flow devices or dipsticks [22], fluorescence polarization immunoassays (FPIAs) [23,24], biosensors and immunochips [25,26] and, more recently, also methods based on the use of aptamers as alternative receptors [27,28]. None of these methods were developed for the simultaneous detection of T-2 and HT-2 and relevant glucosides. A monoclonal antibody was specifically designed and validated for T-2 and T-2G, and when used in ELISA, showed an IC_{50} in the low ng/mL range, suggesting its potential use for their simultaneous detection [29].

Among screening methods, FPIAs have been widely shown to be a useful tool for mycotoxin screening [30,31]. FPIAs are homogeneous immunoassays based on the competition between the analyte and tracer (fluorescent derivative of analyte) for a limited amount of antibody. The analyte content is determined by measuring the reduction of fluorescence polarization value, which is determined by the reduction of tracer molecules able to bind antibody in solution [30]. Several FPIA methods have been developed as screening tools for the determination of major mycotoxins, including aflatoxins, ochratoxin A, zearalenone, fumonisins, deoxynivalenol and T-2 and HT-2, in food matrices [30–32]. In particular, some FPIAs have been developed and validated for the determination of T-2 and HT-2, express as sum, in wheat, oats, barley and cereal-based products [23,24]. To date, no FPIAs and, more generally, no rapid methods are available for the simultaneous determination of T-2 and HT-2 and relevant glucosides.

For this reason, the aim of this study was to develop and validate an FPIA for the simultaneous determination, expressed as sum, of T-2, HT-2, T-2G and HT-2G in wheat. Two different extraction protocols, using organic (methanol/water) and non-organic (water) solvents, were optimized for the FPIA. The two developed methods based on the FPIA and using the two protocols have been validated in-house as quantitative methods, determining sensitivity, recovery and precision values. Furthermore, the two methods were validated through a single-laboratory validation protocol according to harmonized guidelines recently established by the Regulation (EU) No. 519/2014 [33]. The fitness-of-purpose of the FPIA was evaluated by calculating the method precision profiles and setting the screening target concentrations (STC) for false suspect rate and cut-off level to the EU's indicative levels of the sum of T-2 and HT-2 in wheat [33].

2. Results and Discussion

2.1. Development of the FPIA

The antibody specific to the mycotoxin of interest and the tracer are key reagents in the development of a competitive FPIA for mycotoxin analysis. The most important features of the FPIA, such as incubation time, recovery, precision and sensitivity, are strictly related to the antibody/tracer combination used [30]. Binding experiments were performed, in buffer solution, by FP measurement to select the best antibody/tracer combination. Specifically, 48 different combinations were tested, derived from 12 monoclonal antibodies (i.e., ten specifics for T-2G, one for T-2 and one for HT-2) versus 4 different tracers (i.e., one T-2 and three HT-2 fluorescent derivatives). Table 1 reports the maximum values of polarization shift (ΔP_{max}, maximum tracer-antibody binding) and optimized MAb concentrations obtained for each antibody/tracer combination. Among all MAb/tracer combinations tested, the highest bindings were observed for the four combinations obtained with anti-T2 and anti-HT2 versus T2-FL and HT2-FL$_{1a}$, with ΔP_{max} ranging from 205 to 282 mP, and for fifteen combinations composed of anti-T2-glucoside antibodies and the four tracers with ΔP_{max} in the range of 133–280 mP (Table 1).

Table 1. Maximum value of polarization shift (ΔP_{max}) obtained at the optimized antibody concentrations for each antibody/tracer combination.

MAb	Clone	[MAb] (µg/mL)	ΔP_{max} (mP) [1]			
			T2-FL (Dilution 1:3600) [2]	HT2-FL$_{1a}$ (Dilution 1:3000) [2]	HT2-FL$_{1b}$ (Dilution 1:400) [2]	HT2-FL$_2$ (Dilution 1:3600) [2]
Anti-T2G	1–2	190	138 [3]	-	59	57
	1–3	40	280 [3]	39	262 [3]	159 [3]
	1–4	104	225 [3]	28	253 [3]	176 [3]
	2–5	89	116	53	246 [2]	27
	2–11	118	144 [3]	18	50	18
	2–13	134	191 [3]	23	28	19
	2–16	90	200 [3]	37	82	21
	2–17	120	129	27	156 [3]	18
	2–21	132	159 [3]	22	14	16
	2–44	155	176 [3]	133 [3]	246 [3]	49
Anti-T2	1	6	282 [3]	217 [3]	93	111
Anti-HT2	H10-A10	8	205 [3]	230 [3]	122	20

[1] $\Delta P_{max} = mP_{MAb} - mP_{tracer}$; $-\Delta P_{max} < 10$ mP; [2] Optimised dilution (v/v) of the stock solutions providing a total fluorescence intensity equal to 3-fold the blank signal measured for PBS-A; [3] selected antibody/tracer combinations.

All these selected combinations in the optimized concentrations were tested for competitive FPIAs with mixed T-2, HT-2, T-2G and HT-2G (ratio 1:1:0.5:0.5) standard solutions in different ranges of concentrations. Although all T2-glucoside MAbs exhibited suitable cross-reactivity for T-2, HT-2 and HT-2G, the assays showed poor sensitivity, with $IC_{50} \geq 12.2$ ng/mL. The FPIA performed by Anti-T2/T2-FL and Anti-T2/HT2-FL$_{1a}$ showed an acceptable sensitivity, with $IC_{50} \geq 4.0$ ng/mL, but a low cross-reactivity for HT-2, T-2G and HT-2G ranging from 41 to 63%. On the other hand, the FPIA performed by using the Anti-HT2/HT2-FL$_{1a}$ combination showed good sensitivity, with $IC_{50} =$ 2.0 ng/mL and a high cross-reactivity of 100% for T-2 and 80% for T-2G and HT-2G (Figures S1–S4 report the calibration curves for the single toxins). Figure 2 reports the calibration curve for mixed standard solutions of T-2, HT-2 and their glucoside forms in the concentration range of 0.1–73.2 ng/mL (expressed as the sum of the toxins). An incubation time of 5 min was selected as optimal for this competitive immunoassay. The combination of Anti-HT2/HT-2FL$_{1a}$ was then selected as the antibody/tracer combination to use for further development and validation of the assay.

2.2. Testing of Extraction Protocols and Evaluation of Matrix Effects

Two rapid extraction protocols based on the use of organic (methanol/water, 90:10, v/v, Protocol A) and non-organic (water, Protocol B) solvents were tested for the FPIA. Preliminary experiments, based on LC-MS analysis of raw extracts, were performed to assess the reliability in terms of accuracy of the protocols for the simultaneous extraction of T-2, HT-2, T-2G and HT-2G from spiked wheat samples at two levels: 300 and 600 µg/kg (expressed as sum). In the case of Protocol A, mean recoveries ranged from 110 to 119% for single toxins and from 112 to 115% for total toxins (expressed as sum). Relative standard deviations were lower than 7% for all tested toxins. In the case Protocol B, mean recoveries ranged from 88 to 106% for T-2G and HT-2G and from 163 to 192% for HT-2, with RSD lower than 18%. T-2 was not detected in any of the tested replicates at either spiking level. The absence of T2, together with the high recoveries for HT-2, was attributed to de-acetylation processes as a result of cereal carboxylesterases inducing the complete conversion of T-2 into HT-2 in water-based extracts, as previously observed [34]. Moreover, recoveries obtained for total toxins, expressed as sum, were between 88% and 98%, with RSD lower than 9% for both protocols. These results show the potential applicability of the tested extraction protocols for the FPIA, aiming to determine the sum of these toxins.

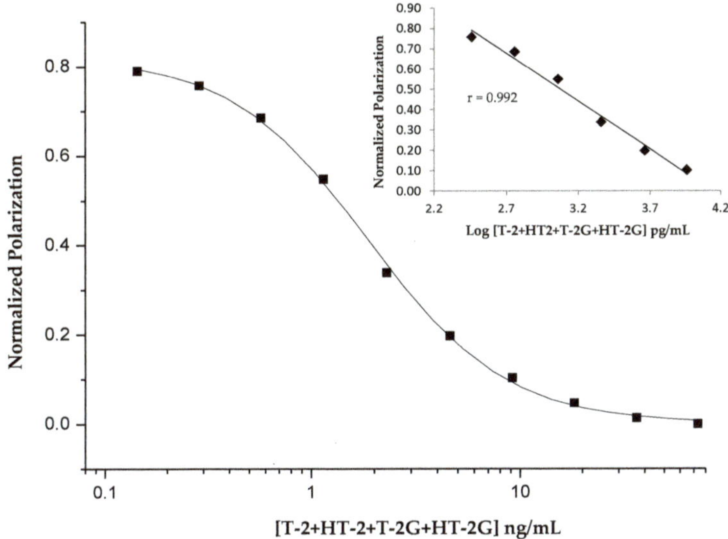

Figure 2. Normalized calibration curve of the selected FPIA obtained with mixed standard solutions of T-2, HT-2, T2-glucoside and HT2-glucoside (expressed as sum, ratio 1:1:0.5:0.5) in PBS-A solution ([Anti-HT2] = 8 µg/mL; [HT2-FL$_{1a}$] obtained after dilution 1:3000 (*v/v*) of the stock solution, see Table 1). The FP linearity range vs. log [T-2 + HT-2 + T-2G + HT-2G] is reported in the insert. Values of the *x*-axes are the toxin concentrations in the final test solution.

Studies to evaluate the presence of matrix interferences by FPIA using different amounts of matrix equivalent (5, 10 and 20 mg) indicated that no significant differences were observed between slopes (t_{calc} < 2.306; p < 0.05) and positions (t_{calc} < 2.262; p < 0.05) of the regression lines obtained with mixed standard solutions and those obtained with spiked diluted extracts for both optimized protocols (Protocol A and Protocol B, calibration curves were reported in Figure S5). These results indicated the total absence, up to the tested amounts, of detectable matrix effects for the developed FPIAs that could produce an overestimation of toxins content.

2.3. Validation of the Methods

To obtain a comprehensive analytical performance profile, the two methods based on the developed FPIA and using the two extraction protocols were validated in-house, both as quantitative methods and as screening methods.

Limit of detection (LOD) and limit of quantification (LOQ) of 10 and 15 µg/kg, respectively, were obtained for the FPIA using both extraction protocols (Protocol A and B) and analyzing 20 mg of matrix equivalent. These results indicated that the sensitivity of the methods was suitable for the quantitative determination of the target toxins at levels far below the indicative level suggested by the European Commission for the sum of T-2 and HT-2 in unprocessed wheat (i.e., 100 µg/kg). Recoveries (%) and repeatability (relative standard deviation, RSD, %) for the FPIA using both extraction protocols from wheat samples spiked with T-2, HT-2, T-2G and HT-2G in the range 50–200 µg/kg (expressed as sum) are reported in Table 2.

Table 2. Average recoveries for T-2, HT-2, T-2G and HT-2G (expressed as sum) and relative standard deviations from spiked wheat obtained by FPIA using protocol A and B.

Spiking Levels (µg/kg)	FPIA			
	Protocol A		Protocol B	
	Recovery	RSD [1] (%)	Recovery	RSD [1] (%)
50	102	13	89	7
100	92	5	98	6
200	96	4	89	6
Overall average	97	9	92	7

[1] RSD, relative standard deviation (*n* = 3 replicates).

Mean recoveries for the FPIA using Protocol A ranged from 92 to 102%, with RSDs lower than 13%, whereas mean recoveries for FPIA using Protocol B were in the range 89–98%, with RSDs lower than 7%. Overall mean recoveries were 97 and 92% for FPIA using Protocols A and B, respectively. The values of recoveries and precision obtained for the developed FPIAs fulfil the criteria of acceptability for an analytical method for the quantitative determination of the native forms fixed by the European Commission [33].

Single-laboratory validation of the two developed methods was performed in-house over 5 different days in accordance with the harmonized guidelines for screening methods established by the Regulation (EU) No. 519/2014 by determining the precision profile of the method, the cut-off level and the false suspect rate. The summary results of the statistical assessment for blank samples and samples artificially contaminated at screening target concentration (STC, 100 µg/kg as the sum of T-2, HT-2, T-2G and HT-2G) are presented in Table 3. The STC value was set at the EU indicative levels of the sum of T-2 and HT-2 in wheat. The mean values of the test responses for the sum of these toxins were 115 and 104 µg/kg for samples at STC and 12 and 21 µg/kg for blank samples, for FPIA with Protocol A and Protocol B, respectively. Depending on the method used, relative standard deviation of the repeatability (RSD_r) and relative standard deviation of intermediate precision (RSD_{RI}) ranged from 5 to 13% for STC samples and from 14 to 25% for blank samples. The calculated cut-off levels were 96 and 80 µg/kg, for FPIA with Protocol A and Protocol B, respectively, and the rate of suspect results for blank samples was in both cases less than 0.1%.

Table 3. Statistical performances of the single-laboratory validation over 5 days of the FPIA for the determination of T-2, HT-2, T-2G and HT-2G (expressed as sum) with blank and artificially contaminated (at the screening target concentration of 100 µg/kg) wheat samples. Cut-off levels and rate of false suspect results were calculated according to the Regulation (EU) No. 519/2014.

Performances	Protocol A		Protocol B	
	Blank	STC [1] (100 µg/kg)	Blank	STC [1] (100 µg/kg)
Mean value [2] (µg/kg)	12	115	21	104
RSD_r [3] (%)	16	5	14	9
RSD_{RI} [4] (%)	25	10	16	13
Cut-off level		96		80
Rate of false suspect results (%)	<0.1		<0.1	

[1] STC, screening target concentration; [2] The mean value of the total content of T-2, HT-2, T-2G and HT-2G (µg/kg, expressed as sum) (*n* = 20 replicates); [3] RSD_r, relative standard deviation of the repeatability; [4] RSD_{RI}, relative standard deviation (intermediate precision).

Moreover, Figure 3 shows the graphical representation of the results, reporting the toxin contents obtained for the 20 artificially contaminated samples at STC and the 20 blanks, analyzed over 5 different days together with the calculated cut-off levels for the FPIA using Protocol A (Figure 3a) and Protocol B

(Figure 3b). A complete separation between blanks and spiked samples at STC was observed for both methods, demonstrating their ability to discriminate between these two groups of samples. This aspect was also confirmed by the low rate of suspect results (less than 0.1%). The overall results confirmed the applicability of both methods for the simultaneous determination of T-2, HT-2, T-2G and HT-2G (expressed as sum) in wheat samples at the indicative level reported at EU level for the native forms T-2 and HT-2 from uncontaminated samples.

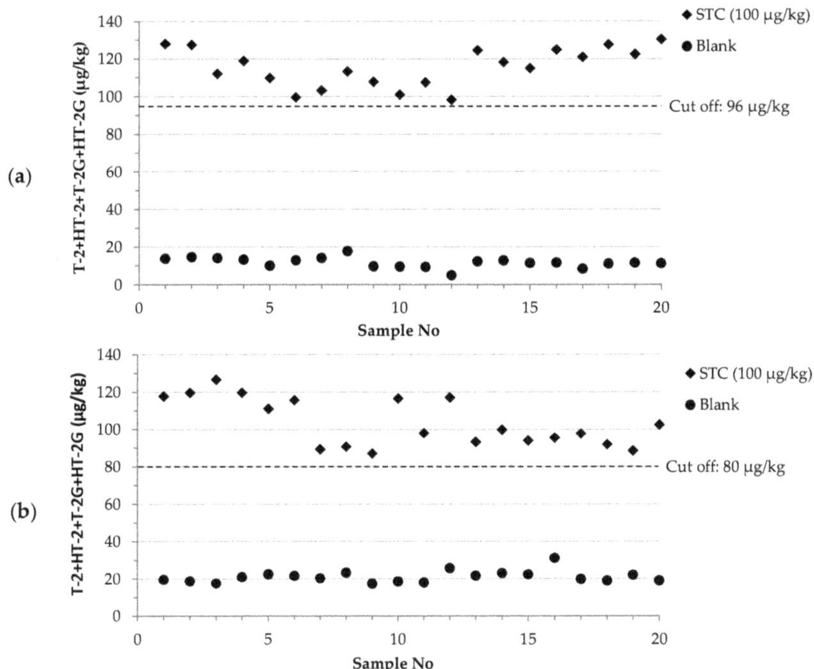

Figure 3. Contents of T-2, HT-2, T-2G and HT-2G (expressed as sum) of 20 artificially contaminated wheat samples at STC (100 µg/kg) and of 20 blank wheat samples analyzed under repeatability conditions on 5 different days: (**a**) FPIA using protocol A; (**b**) FPIA using protocol B.

Moreover, there are several advantages to using a solvent-free extraction (Protocol B). Removal of organic solvents from the extraction is safer, because it reduces exposure of the analyst. It is also less expensive, because the cost associated with management of solvent waste (such as storage and disposal) are eliminated. Finally, eliminating solvents facilitates use in a wider variety of settings outside of the traditional laboratory.

3. Conclusions

A rapid FPIA for the simultaneous determination of T-2, HT-2, T-2G and HT-2G, expressed as their sum, was developed for the first time, by testing 12 monoclonal antibodies and 4 tracers. Two alternative extraction protocols, using organic and non-organic solvents, were optimised for the quantitative determination of these toxins in wheat. The main advantages of the developed methods were ease of use and speed (total time less than 15 min). The main disadvantage was the inability to determine the content of the individual toxins. The methods when validated as quantitative methods showed analytical performances, in terms of recovery (92–97%) and precision (RSDs ≤ 13%) that fulfilled the EU criteria for acceptability of an analytical method for the determination of native forms. Furthermore, the developed methods were also validated according to EU harmonized guidelines for the single-laboratory validation of screening methods. The satisfactory analytical performances, in terms of precision under repeatability (5–16%), intermediate precision (10–25%), cut-off level (80 and 96 µg/kg for the two methods) and rate of false positives (<0.1%), confirmed the applicability of the proposed methods based on the FPIA as screening methods for assessing the content of T2, HT2, T2G and HT2G in wheat at EU indicative levels reported for the native forms. Additional advantages of the FPIA methods were low-cost, portability, amenability to automation, and the use of environmentally friendly extraction procedures. These advantages make the developed FPIA methods useful and robust tools for high-throughput screening of these toxins in wheat, without the need for well equipped laboratories or personnel with a high level of technical skill.

4. Materials and Methods

4.1. Chemicals and Reagents

Acetonitrile and methanol were reagent grade or better and were purchased from Carlo Erba Reagents (Milan, Italy). Ultrapure water was produced by a Milli-Q® Direct system (Merck KGaA, Darmstadt, Germany). T-2 and HT-2, sodium chloride (NaCl), sodium azide (NaN$_3$), phosphate buffer saline (PBS) and ovalbumin (OVA) were purchased from Sigma-Aldrich (Milan, Italy). Ten monoclonal antibodies (MAbs) specific for T-2G (clones 1–2, 1–3, 1–4, 2–5, 2–11, 2–13, 2–16, 2–17, 2–21 and 2–44) were produced by US Department of Agriculture-Agricultural Research Center (USDA-ARS, Peoria, IL, USA) and have been described in Maragos et al. [29]. A Mab specific for T-2 (clone 1) was produced by Chinese Academy of Agricultural Sciences—Oil Crops Research Institute (Wuhan, China) and has been described by Zhang et al. [35]. A Mab specific for HT-2 (clone H10-A10) was purchased from University of Natural Resources and Life Science of Vienna—Department for Agrobiotechnology IFA-Tulln (Tulln, Austria). T-2 glucoside and HT-2 glucoside, both as α-anomers, were produced by USDA-ARS and have been described by McCormick et al. [36]. Glass culture tubes (10 × 75 mm) were obtained from VWR International (Milan, Italy). Paper filters (No. 4) and glass microfiber filters (GF/A) were purchased from Whatman (Maidstone, UK).

4.2. Preparation of Immunoassay Reagent Solutions

T-2 and HT-2 stock solutions, at the concentration of 1 mg/mL, were prepared by dissolving solid commercial toxins in acetonitrile. T-2G and HT-2G stock solutions were prepared at the concentration of 1 mg/mL in acetonitrile. Diluted T-2, HT-2, T-2G and HT-2G solutions were prepared in acetonitrile at the concentration of 20 µg/mL. Mixed standard solutions of T-2, HT-2, T-2G and HT-2G (ratio 1:1:0.5:0.5) were prepared in acetonitrile and PBS-A (PBS, 10 mM, pH = 7.4, containing 0.1% of NaN$_3$) at the concentration of 10 and 3 µg/mL, respectively. The mixed standard solution in acetonitrile was used for spiking experiments in the LC-MS analysis to evaluate the extraction efficiency of the two alternative protocols. The mixed standard solution in PBS-A was used to prepare FPIA calibration curves. T-2 and HT-2 tracers (T2-FL: fluorescein-labelled T2 toxin; HT2-FL$_{1a}$ and HT2-FL$_{1b}$: monosubstituted fluorescein-labelled HT2 toxin, arbitrarily ascribed to each isomeric product; HT2-FL$_2$: bi-substituted fluorescein-labelled HT2 toxin tracers) were prepared according to the procedure reported by Lippolis et al. 2011 [23], who also reported their chemical structures. Tracer working solutions were prepared daily by diluting the relevant stock solutions in methanol at the concentration providing a total fluorescence intensity equal to 3-fold the blank signal measured for the assay buffer (PBS-A). In the case of the selected tracer (HT2-FL$_{1a}$) for the developed FPIA the dilution ratio was 1:3000 (*v/v*).The twelve monoclonal antibodies were diluted with PBS-OVA (PBS-A, containing 0.1% of OVA) according to the experiments reported in the Section 4.4 (FPIA analysis).

4.3. Sample Preparations

Durum wheat samples of different cultivars were collected from several fields in Italy. Samples were milled by the Ultra Centrifugal Mill ZM 200 (Retsch Technology GmbH, Hann, Germany) equipped with a 500-µm sieve. Two different extraction protocols, using organic (methanol/water) and non-organic (water) solvents, were tested. In particular, in the first approach (Protocol A), an aliquot (50 g) of wheat, added together with NaCl (1 g), was extracted with 100 mL of methanol/water (90/10, *v/v*) by blending at high speed for 3 min using a Steril Mixer 12 blender (VWR International). After filtering extracts through the filter paper, they were diluted with a 4% NaCl solution (ratio 1:5, *v/v*) and left to rest for 5 min. The dilution with 4% NaCl solution was carried out in order to reduce the percentage of methanol in solution and to precipitate interfering compounds, which may contribute to increasing the matrix effect. The diluted extracts were filtered by glass microfiber filters and analyzed by FPIA. In the case of the second approach (Protocol B), an aliquot of ground wheat samples (10 g) was extracted with 100 mL of water by blending at high speed for 3 min by Steril Mixer 12 blender. To evaluate the performances of both protocols for the simultaneous extraction of the tested toxins, recovery experiments were performed in triplicate by spiking uncontaminated (blank) wheat samples with a mixed T-2, HT-2, T-2G and HT-2G (ratio 1:1:0.5:0.5) spiking solution in acetonitrile at levels of 300 and 600 µg/kg. Extracts of both protocols (A and B) were filtered through a filter paper and a glass microfiber filter and subsequently analyzed by LC-MS analysis according to the experimental conditions reported in Lattanzio et al. 2012 [8].

4.4. FPIA Analysis

All FP measurements were carried out in the glass culture tubes and using a portable reader (Sentry® 100, Diachemix Corporation, Milwaukee, WI, USA) with excitation and emission wavelengths of 485 and 535, respectively. Preliminary FP measurements were performed in buffer solution in order to find the best combination antibody/tracer to be used in the FPIA. In particular, the binding ability of the 12 MAbs was tested versus the 4 synthesized tracers by measuring the polarization shift ($\Delta P = mP_{MAb} - mP_{tracer}$) observed between the test PBS-A solution containing the tracer (at the optimized concentration, see Table 1) and the test solution after adding the MAb working solutions at different concentrations and incubating in the range 0–10 min. In particular, the MAb working solution were 47.5–190 µg/mL for clone 1–2, 3.95–79 µg/mL for clone 1–3, 5.2–104 µg/mL for clone 1–4, 17.8–89 µg/mL for clone 2–5, 29.5–118 µg/mL for clone 2–11, 33.5–134 µg/mL for clone 2–13, 18.0–90 µg/mL for clone 2–16, 30.0–120 µg/mL for clone 2–17, 33.0–132 µg/mL for clone 2–21, 15.5–155 µg/mL for clone 2–44, 0.20–200 µg/mL for anti-T2 clone 1, 1.48–39.4 µg/mL for anti-HT2 clone H10-A10. For each Mab/tracer combination, the optimized MAb concentration corresponded to the lowest concentration providing the maximum value of ΔP (ΔP_{max}).Competitive FPIAs were carried out with mixed (ratio 1:1:0.5:0.5) standard solutions, in different concentration ranges, by using the nineteen selected antibody/tracer combinations (indicated in Table 1, using an arbitrary cut-off for the ΔP_{max} of 130 mP). In particular, the assays were carried out by adding 850 µL of PBS-A, 100 µL of antibody working solution and 50 µL of mixed T-2, HT-2, T-2G and HT-2G standard solution in a test tube. The polarization value of this test solution, previously mixed by vortex, was measured and used as the blank. An aliquot (25 µL) of tracer working solution (at the optimized concentration, see Table 1) was added and the solution was gently mixed by vortex. After incubating (in the range 0–10 min) the polarization value, expressed in millipolarization units (mP), of the solution was measured. Competitive FPIAs were also performed using the selected antibody/tracer combination (clone H10-A10/HT2-FL$_{1a}$) and standard solutions of the single toxins T-2, HT-2, T-2G and HT-2G to determine their midpoint concentrations (IC_{50}).The cross-reactivity of the selected monoclonal antibody (clone H10-A10) has already been tested for structurally related toxins (deoxynivalenol, 3-acetyl-DON, 15-acetyl-DON, diacetoxyscirpenol, neosolaniol and nivalenol) and mycotoxins frequently occurring in wheat (ochratoxin A and zearalenone), showing a very low cross-reactivity for neosolaniol (CR% = 0.12%) and no cross-reactivity for all other tested toxins [23].

The developed FPIA were carried out by adding and mixing 700 µL of PBS-A, 100 µL of antibody working solution (clone H10-A10, 8 µg/mL), 200 µL of filtered extract for both protocols (Protocol A and Protocol B, equivalent to 20 mg of matrix) or 50 µL of mixed T-2, HT-2, T-2G and HT-2G standard solution. After reading the blank using an FP reader, 25 μL of tracer working solution (HT2-FL$_{1a}$, dilution 1:3000, *v/v* of the stock solution) was added, and the final solution was mixed and incubated for 5 min. The polarization value after incubation was measured. The measured polarization values were normalized to fit in the range 0–1, the equation $Y_{obs} = (mP_{obs} - mP_0)/(mP_1 - mP_0)$ was used, where mP_{obs}, mP_0 and mP_1 are the polarization of the test solution, of an antibody-free control solution and of a toxin-free control solution, respectively, and Y_{obs} is the normalized result for the test solution [30]. The content of T-2, HT-2, T-2G and HT-2G, expressed as sum, in the wheat extracts was determined by using the measured normalized polarization values and the FPIA calibration curves in the toxins concentration range 5.85–187.5 ng/mL.

4.5. Evaluation of Matrix Effects

The presence of matrix effect on the developed FPIA was evaluated for both protocols (Protocol A and Protocol B). In particular, diluted extracts of blank wheat were spiked at different T-2, HT-2, T-2G and HT-2G levels in the concentration range 5.85–187.5 ng/mL (expressed as sum) at different amount of matrix equivalent analyzed, i.e., 5, 10 and 20 mg. Calibration curves determined by using either standard solutions or spiked diluted extract of uncontaminated wheat samples were compared.

4.6. Validation as Quantitative Methods

The developed methods were validated in-house as quantitative methods, with their performances being evaluated in terms of sensitivity, recovery and repeatability. Experiments were carried out to determine the sensitivity of the FPIA using both protocols (A and B). In particular, limits of detection (LODs) of the FPIAs were calculated from the mean FP signals of representative uncontaminated wheat samples ($n = 10$, wheat samples of different cultivars) minus 3 standard deviations of the mean signal. Limits of quantification (LOQs) were calculated by determining the lowest amount of measured toxins (expressed as sum) that was quantitatively determined by the calibration curve within the linearity range of the FPIA.

Recovery experiments were performed by spiking, in triplicate, blank wheat samples at levels of 50, 100 and 200 μg/kg and subsequently analyzing them by FPIA using both optimized extraction protocols (Protocol A and Protocol B). This contamination range was selected in order to include the indicative level provided by the European Union for T-2 and HT-2 (100 μg/kg). Samples were left overnight at room temperature to allow solvent evaporation prior to the FPIA analysis.

4.7. Validation as Screening Methods

The developed methods were also validated through single-laboratory validation according to harmonized guidelines for screening methods established by the European Commission in the Regulation (EU) No. 519/2014 [33]. The fitness-of-purpose of the FPIAs was evaluated by calculating the method precision profiles and setting the screening target concentrations [33] for the false suspect rate and the cut-off level to the EU's regulatory/indicative levels of the relevant native forms in durum wheat. The validation design required 2 sample sets: (1) 20 positive control samples, namely, wheat samples fortified with the mycotoxins at the STC (100 μg/kg for the sum of T-2, HT-2, T-2G and HT-2G), and (2) 20 negative samples (blank wheat samples). The measurements for each validation level (STC and blank) were evenly distributed over 5 different days, resulting in 4 independent analyses per day. Cut-off levels were calculated from the results obtained for the 20 samples spiked at the STC by using the equation provided in the Regulation (EU) No. 519/2014:

$$\text{Cut-off} = R_{STC} - t\text{-value}_{(0.05)} \times SD_{STC} \tag{1}$$

where R_{STC} is the mean response of the spiked samples at STC; t-value$_{(0.05)}$ is the one tailed t-value for a rate of false negative results of 5% (i.e., 1.729 for 19 degrees of freedom and 20 replicates); the SD$_{STC}$ is the standard deviation of intermediate precision (which is the sum of repeatability and between-day variability) at STC calculated by one-way ANOVA (p-value = 0.05). The false suspect rate is determined on the base of the results obtained for 20 blanks and the calculated cut-off by calculating the t-value from the following equation:

$$t\text{-value} = (\text{mean}_{\text{blank}} - \text{cut-off})/\text{SD}_{\text{blank}} \qquad (2)$$

where $\text{mean}_{\text{blank}}$ is the mean response of the 20 blank samples; SD_{blank} is the standard deviation of the intermediate precision of blank samples. From the calculated t-value, the rate of false suspect results for a one-tailed distribution was calculated as indicated in Section 4.8.

4.8. Statistical Analysis

FPIA data were fit to linear or sigmoidal equations with Origin software version 6.0 (OriginLab Corporation, Northampton, MA, USA, 1999) using the unweighted least-square method. For the sigmoidal fit, the equation used was of the form $y = A_2 + [A_1 - A_2/1 + (x/x_0)^P]$. Here, A_1 and A_2 represent the initial value (left horizontal asymptote) and the final value (right horizontal asymptote), respectively, while x_0 represents the inflection point (center), and P represents the power. In the experiments to measure matrix effects, the linear regression curves were compared using parallelism and position statistical tests [37]. In the recovery experiments, which used three spiking levels, the homogeneities of the variances and the homogeneities of the means were compared using Barlett's test and one-way ANOVA (p-value = 0.05), respectively. Results from experiments from the single-laboratory validation studies were subjected to one-way ANOVA (p-value = 0.05) using the Microsoft Excel add-on. The rate of false suspect results relative to the calculated t-value from a one-tailed Student's T Distribution was obtained using the spreadsheet function "TDIST".

Supplementary Materials: The following are available online at http://www.mdpi.com/2072-6651/11/7/380/s1, Figure S1: Normalized calibration curve of the selected FPIA obtained with mixed standard solutions of T-2 toxin (T-2) in PBS-A solution ([Anti-HT2] = 8 µg/mL; [HT2-FL$_{1a}$] obtained after dilution 1:3000 (v/v) of the stock solution). Figure S2: Normalized calibration curve of the selected FPIA obtained with mixed standard solutions of HT-2 toxin (HT-2) in PBS-A solution ([Anti-HT2] = 8 µg/mL; [HT2-FL$_{1a}$] obtained after dilution 1:3000 (v/v) of the stock solution). Figure S3: Normalized calibration curve of the selected FPIA obtained with mixed standard solutions of T-2 glucoside (T-2G) in PBS-A solution ([Anti-HT2] = 8 µg/mL; [HT2-FL$_{1a}$] obtained after dilution 1:3000 (v/v) of the stock solution). Figure S4: Normalized calibration curve of the selected FPIA obtained with mixed standard solutions of HT-2 glucoside (HT-2G) toxin in PBS-A solution ([Anti-HT2] = 8 µg/mL; [HT2-FL$_{1a}$] obtained after dilution 1:3000 (v/v) of the stock solution). Figure S5: Calibration curves (concentration range from 0.3 to 9.1 ng/mL) obtained with mixed standard solutions of T-2, HT-2, T2-glucoside and HT2-glucoside (expressed as sum, ratio 1:1:0.5:0.5) (black square) and spiked diluted extracts of wheat, obtained using Protocol A (a) and Protocol B (b), by analyzing 5 mg (multiplication sign), 10 mg (white triangle) and 20 mg (black circle) of matrix equivalent.

Author Contributions: Conceptualization, V.L., C.M.M., A.F.L. and M.P.; Formal analysis, A.C.R.P., E.M., B.C. and V.M.T.L.; Methodology, V.L., A.C.R.P., E.M., C.M.M., S.M., P.L. and M.P.; Supervision, M.P.; Validation, V.M.T.L.; and A.D.G.; Writing—original draft, V.L.; Writing—review & editing, V.L.; V.M.T.L.; A.D.G.; C.M.M., S.M., P.L., A.F.L. and M.P.

Funding: This work has been supported by the MYCOKEY project "Integrated and innovative key actions for mycotoxin management in the food and feed chain" (H2020-Grant Agreement No. 678781).

Conflicts of Interest: All authors declare no conflict of interest.

Disclaimer: Mention of trade names or commercial products in this publication is solely for the purpose of providing specific information and does not imply recommendation or endorsement by the U.S. Department of Agriculture. USDA is an equal opportunity provider and employer.

References

1. Canady, R.A.; Coker, R.D.; Egan, S.K.; Krska, R.; Olsen, M.; Resnik, S.; Schlatter, J. T-2 and HT-2 toxins. In *Safety Evaluation of Certain Mycotoxins in Food*; WHO/FAO, Ed.; WHO Food Additives Series 47, FAO Food and Nutrition Paper 74; WHO: Geneva, Switzerland, 2001; pp. 557–638. Available online: http://www.fao.org/3/a-bc528e.pdf (accessed on 24 April 2019).

2.   Schothorst, R.C.; van Egmond, H.P. Report from SCOOP task 3.2.10 "Collection of occurrence data of Fusarium toxins in food assessment of dietary intake by the population of EU member states" subtask: Trichothecenes. *Toxicol. Lett.* **2004**, *153*, 133–153. [CrossRef] [PubMed]

3.   European Food Safety Authority (EFSA). Scientific opinion on the risks for animal and public health related to the presence of T-2 and HT-2 toxin in food and feed. *EFSA J.* **2011**, *9*, 2481. [CrossRef]

4.   European Food Safety Authority (EFSA). Human and animal dietary exposure to T-2 and HT-2 toxin. *EFSA J.* **2017**, *15*, 4972.

5.   Pettersson, H. Toxicity and risks with T-2 and HT-2 toxins in cereals. *Plant Breed. Seed Sci.* **2011**, *64*, 65–74. [CrossRef]

6.   European Commission (EC). Commission recommendation of 27 March 2013 on the presence of T-2 and HT-2 toxin in cereals and cereal products (2013/165/EU). *Off. J. Eur. Union* **2013**, *91*, 12–15.

7.   European Food Safety Authority (EFSA). Appropriateness to set a group health based guidance value for T2 and HT2 toxin and its modified forms. *EFSA J.* **2017**, *15*, 4655.

8.   Lattanzio, V.M.T.; Visconti, A.; Haidukowski, M.; Pascale, M. Identification and characterization of new Fusarium masked mycotoxins, T2 and HT2 glycosyl derivatives, in naturally contaminated wheat and oats by liquid chromatography-high-resolution mass spectrometry. *J. Mass Spectrom.* **2012**, *47*, 466–475. [CrossRef] [PubMed]

9.   Veprikova, Z.; Vaclavikova, M.; Lacina, O.; Dzuman, Z.; Zachariasova, M.; Hajslova, J. Occurrence of mono- and di-glycosylated conjugates of T-2 and HT-2 toxins in naturally contaminated cereals. *World Mycotoxin J.* **2012**, *5*, 231–240. [CrossRef]

10.   Nakagawa, H.; Sakamoto, S.; Sago, Y.; Kushiro, M.; Nagashima, H. The use of LC-Orbitrap MS for the detection of Fusarium masked mycotoxins: The case of type A trichothecenes. *World Mycotoxin J.* **2012**, *5*, 271–280. [CrossRef]

11.   Lattanzio, V.M.T.; Ciasca, B.; Terzi, V.; Ghizzoni, R.; McCormick, S.P.; Pascale, M. Study of the natural occurrence of T-2 and HT-2 toxins and their glucosyl derivatives from field barley to malt by high resolution Orbitrap mass spectrometry. *Food Addit. Contam. Part A* **2015**, *32*, 1647–1655. [CrossRef]

12.   Bryla, M.; Waskiewicz, A.; Ksieniewicz-Wozniak, E.; Szymczyk, K.; Jedrzejczak, R. Modified Fusarium mycotoxins in cereals and their products-metabolism, occurrence, and toxicity: An updated review. *Molecules* **2018**, *23*, 963. [CrossRef] [PubMed]

13.   Meneely, J.P.; Ricci, F.; van Egmond, H.P.; Elliott, C.T. Current methods of analysis for the determination of trichothecene mycotoxins in food. *TrAC Trends Anal. Chem.* **2011**, *30*, 192–203. [CrossRef]

14.   Pascale, M.; Panzarini, G.; Visconti, A. Determination of HT-2 and T-2 toxins in oats and wheat by ultra-performance liquid chromatography with photodiode array detection. *Talanta* **2012**, *89*, 231–236. [CrossRef] [PubMed]

15.   Krska, R.; Malachova, A.; Berthiller, F.; Van Egmond, H.P. Determination of T-2 and HT-2 toxins in food and feed: An update. *World Mycotoxin J.* **2014**, *7*, 131–142. [CrossRef]

16.   Nathanail, A.V.; Syvähuoko, J.; Malachová, A.; Jestoi, M.; Varga, E.; Michlmayr, H.; Adam, G.; Sieviläinen, E.; Berthiller, F.; Peltonen, K. Simultaneous determination of major type A and B trichothecenes, zearalenone and certain modified metabolites in Finnish cereal grains with a novel liquid chromatography-tandem mass spectrometric method. *Anal. Bioanal. Chem.* **2015**, *407*, 4745–4755. [CrossRef]

17.   Leslie, J.F.; Lattanzio, V.; Audenaert, K.; Battilani, P.; Cary, J.; Chulze, S.N.; De Saeger, S.; Gerardino, A.; Karlovsky, P.; Liao, Y.-C.; et al. Mycokey round table discussions on future directions in research on chemical detection methods, genetics and biodiversity of mycotoxins. *Toxins* **2018**, *10*, 109. [CrossRef] [PubMed]

18.   Peng, D.; Chang, F.; Wang, Y.; Chen, D.; Liu, Z.; Zhou, X.; Feng, L.; Yuan, Z. Development of a sensitive monoclonal-based enzyme-linkedimmunosorbentassay for monitoring T-2 toxin in food and feed. *Food Addit. Contam. A* **2016**, *33*, 683–692. [CrossRef]

19.   Li, Y.; Luo, X.; Yang, S.; Cao, X.; Wang, Z.; Shi, W.; Zhang, S. High specific monoclonal antibody production and development of an ELISA method for monitoring T-2 toxin in rice. *J. Agric. Food Chem.* **2014**, *62*, 1492–1497. [CrossRef]

20.   Wang, J.; Duan, S.; Zhang, Y.; Wang, S. Enzyme-linked immunosorbent assay for the determination of T-2 toxin in cereals and feedstuff. *Microchim. Acta* **2010**, *169*, 137–144. [CrossRef]

21. Yoshizawa, T.; Kohno, H.; Ikeda, K.; Shinoda, T.; Yokohama, H.; Morita, K.; Kusada, O.; Kobayashi, Y. A practical method for measuring deoxynivalenol, nivalenol, and T-2 + HT-2 toxin in foods by an enzyme-linked immunosorbent assay using monoclonal antibodies. *Biosci. Biotechnol. Biochem.* **2004**, *68*, 2076–2085. [CrossRef]

22. Molinelli, A.; Grossalber, K.; Führer, M.; Baumgartner, S.; Sulyok, M.; Krska, R. Development of qualitative and semiquantitative immunoassay-based rapid strip tests for the detection of T-2 toxin in wheat and oat. *J. Agric. Food Chem.* **2008**, *56*, 2589–2594. [CrossRef] [PubMed]

23. Lippolis, V.; Pascale, M.; Valenzano, S.; Pluchinotta, V.; Baumgartner, S.; Krska, R.; Visconti, A. A rapid fluorescence polarization immunoassay for the determination of T-2 and HT-2 toxins in wheat. *Anal. Bioanal. Chem.* **2011**, *401*, 2561–2571. [CrossRef] [PubMed]

24. Porricelli, A.C.R.; Lippolis, L.; Valenzano, S.; Cortese, M.; Suman, M.; Zanardi, S.; Pascale, M. Optimization and validation of a fluorescence polarization immunoassay for rapid detection of T-2 and HT-2 toxins in cereals and cereal-based products. *Food Anal. Methods* **2016**, *9*, 3310–3318. [CrossRef]

25. Meneely, J.P.; Sulyok, M.; Baumgartner, S.; Krska, R.; Elliott, C.T. A rapid optical immunoassay for the screening of T-2 and HT-2 toxin in cereals and maize-based baby food. *Talanta* **2010**, *81*, 630–636. [CrossRef] [PubMed]

26. Wang, Y.; Liu, N.; Ning, B.; Liu, M.; Lv, Z.; Sun, Z.; Peng, Y.; Chen, C.; Li, J.; Gao, Z. Simultaneous and rapid detection of six different mycotoxins using an immunochip. *Biosens. Bioelectron.* **2012**, *34*, 44–50. [CrossRef] [PubMed]

27. Khan, I.M.; Zhao, S.; Niazi, S.; Mohsin, A.; Shoaib, M.; Duan, N.; Wu, S.; Wang, Z. Silver nanoclusters based FRET aptasensor for sensitive and selective fluorescent detection of T-2 toxin. *Sensor. Actuators B Chem.* **2018**, *277*, 328–335. [CrossRef]

28. Zhang, M.; Wang, Y.; Yuan, S.; Sun, X.; Huo, B.; Bai, J.; Peng, Y.; Ning, B.; Liu, B.; Gao, Z. Competitive fluorometric assay for the food toxin T-2 by using DNA-modified silver nanoclusters, aptamer-modified magnetic beads, and exponential isothermal amplification. *Microchim. Acta* **2019**, *186*, 219. [CrossRef]

29. Maragos, C.M.; Kurtzman, C.; Busman, M.; Price, N.; McCormick, S. Development and evaluation of monoclonal antibodies for the Glucoside of T-2 Toxin (T2-Glc). *Toxins* **2013**, *5*, 1299–1313. [CrossRef] [PubMed]

30. Lippolis, V.; Maragos, C. Fluorescence polarization immunoassays for rapid, accurate and sensitive determination of mycotoxins. *World Mycotoxin J.* **2014**, *7*, 479–489. [CrossRef]

31. Zhang, H.; Yang, S.; De Ruyck, K.; Beloglazova, N.; Eremin, S.A.; De Saeger, S.; Zhang, S.; Shen, J.; Wang, Z. Fluorescence polarization assays for chemical contaminants in food and environmental analyses. *TrAC Trends Anal. Chem.* **2019**, *114*, 293–313. [CrossRef]

32. Lippolis, V.; Porricelli, A.C.R.; Cortese, M.; Suman, M.; Zanardi, S.; Pascale, M. Determination of ochratoxin A in Rye and Rye-Based products by Fluorescence polarization immunoassay. *Toxins* **2017**, *9*, 305. [CrossRef] [PubMed]

33. European Commission (EC). Commission Regulation (EU) 519/2014 of 16 May 2014 amending Regulation (EC) No 401/2006 as regards methods of sampling of large lots, spices and food supplements, performance criteria for T-2, HT-2 toxin and citrinin and screening methods of analysis. *Off. J. Eur. Union* **2014**, *147*, 29–43.

34. Lattanzio, V.M.T.; Solfrizzo, M.; Visconti, A. Enzymatic hydrolysis of T-2 toxin for the quantitative determination of total T-2 and HT-2 toxins in cereals. *Anal. Bioanal. Chem.* **2009**, *395*, 1325–1334. [CrossRef] [PubMed]

35. Zhang, Z.; Wang, D.; Li, J.; Zhang, Q.; Li, P. Monoclonal antibody–europium conjugate-based lateral flow time-resolved fluoroimmunoassay for quantitative determination of T-2 toxin in cereals and feed. *Anal. Methods* **2015**, *7*, 2822–2829. [CrossRef]

36. McCormick, S.P.; Price, N.P.J.; Kurtzman, C.P. Glucosylation and other biotransformations of T-2 toxin by yeasts of the *Trichomonascus Clade*. *Appl. Environ. Microbiol.* **2012**, *78*, 8694–8702. [CrossRef] [PubMed]

37. Soliani, L. Statistica applicata per la ricerca e professioni scientifiche. In *Manuale di Statistica Univariata e Bivariata Parametrica e Non-Parametrica*; Uninova-Gruppo Pegaso: Parma, Italy, 2007.

Article

Emerging Fusarium Mycotoxins Fusaproliferin, Beauvericin, Enniatins, and Moniliformin in Serbian Maize

Igor Jajić [1], Tatjana Dudaš [1,*], Saša Krstović [1], Rudolf Krska [2,3], Michael Sulyok [2], Ferenc Bagi [1], Zagorka Savić [1], Darko Guljaš [1] and Aleksandra Stankov [1]

1   Faculty of Agriculture, University of Novi Sad, 21000 Novi Sad, Serbia; igor.jajic@stocarstvo.edu.rs (I.J.); sasa.krstovic@stocarstvo.edu.rs (S.K.); bagifer@polj.uns.ac.rs (F.B.); zagorka.savic@polj.uns.ac.rs (Z.S.); darko.guljas@stocarstvo.edu.rs (D.G.); aleksandra.stankov@polj.uns.ac.rs (A.S.)

2   Institute of Bioanalytics and Agro-Metabolomics, Department IFA-Tulln, University of Natural Resources and Life Sciences Vienna (BOKU), A-3430 Tulln, Austria; rudolf.krska@boku.ac.at (R.K.); michael.sulyok@boku.ac.at (M.S.)

3   Institute for Global Food Security, School of Biological Sciences, Queens University Belfast, University Road, Belfast BT7 1NN, UK

*   Correspondence: tatjana.dudas@polj.uns.ac.rs; Tel.: +381-655-237-432-54

Received: 30 March 2019; Accepted: 17 June 2019; Published: 19 June 2019

Abstract: Emerging mycotoxins such as moniliformin (MON), enniatins (ENs), beauvericin (BEA), and fusaproliferin (FUS) may contaminate maize and negatively influence the yield and quality of grain. The aim of this study was to determine the content of emerging *Fusarium* mycotoxins in Serbian maize from the 2016, 2017, and 2018 harvests. A total of 190 samples from commercial maize production operations in Serbia were analyzed for the presence of MON, ENs, BEA, and FUS using liquid chromatography-tandem mass spectrometry (LC-MS/MS). The obtained results were interpreted together with weather data from each year. MON, BEA, and FUS were major contaminants, while other emerging mycotoxins were not detected or were found in fewer samples (<20%). Overall contamination was highest in 2016 when MON and BEA were found in 50–80% of samples. In 2017 and 2018, high levels of MON, FUS, and BEA were detected in regions with high precipitation and warm weather during the silking phase of maize (July and the beginning of August), when the plants are most susceptible to *Fusarium* infections. Since environmental conditions in Serbia are favorable for the occurrence of mycotoxigenic fungi, monitoring *Fusarium* toxins is essential for the production of safe food and feed.

Keywords: emerging mycotoxins; *Fusarium*; LC-MS/MS; maize; Serbia

Key Contribution: Up to present, little effort has been put forth in terms of quality control regarding emerging mycotoxins in Serbian maize. This article gives the first insight into the current situation of emerging mycotoxin contaminations in Serbian maize and recommends further monitoring studies necessary for legislative purposes since climate conditions are favorable for the development of mycotoxin-producing fungi that pose a threat to food and food quality.

---

1. Introduction

Mycotoxins have become one of the most important food contaminants in modern society. They are toxic secondary metabolites that are usually produced by *Aspergillus*, *Penicillium*, and *Fusarium* fungi in favorable environmental conditions. Among these species, *Fusarium* are the most prevalent mycotoxin-producing fungi in the northern temperate regions, and mainly Central and SoutheasternEurope [1]. *Fusarium* molds are known as producers of several mycotoxins, including both "traditional" as trichothecenes, zearalenone, and the fumonisins [2] and "emerging toxins" as moniliformin (MON), enniatins (ENs), beauvericin (BEA), and fusaproliferin (FUS) [3]. Additionally, Kovalsky et al. [3] found a co-occurrence of "traditional" toxins with EN, MON, and BEA. The main producers of emerging *Fusarium* mycotoxins in cereals are *Fusarium avenacum, Fusarium verticillioides (moniliforme), Fusarium proliferatum,* and *Fusarium subglutinans* [1].

MON is a sodium or potassium salt of 1-hydroxycyclobut-1-ene-3,4-dione [4]. It was discovered by Cole et al. [5] while screening for toxigenic products of *F. verticillioides* isolated from southern leaf blight-damaged maize seed. BEA and ENs belong to a group of cyclodepsipeptidesthat may have antibiotic, insecticidal, and cytotoxic effects [6–9]. FUS is a toxic sesterterpene originally isolated by Ritieni et al. [10] from *F. proliferatum* in autoclaved maize cultures.

BEA is known as a cholesterol acyltransferase inhibitor [11] and it is toxic to several human cell lines [12,13]. Additionally, BEA can induce apoptosis and DNA fragmentation [14]. FUS is a sesterterpene identified from maize cultures of *F. proliferatum* isolated from maize [15]. FUS is toxic to *Artemiasalina*, to the lepidopteran cell line SF-9 and to the human nonneoplastic B-lymphocyte cell line IARC/LCL 171 [16]. FUS can induce teratogenic effects in chicken embryos [17]. Kriek et al. [18] found that ducklings and rats fed diets containing MON led to muscular weakness, respiratory distress, cyanosis, coma, and death.

There is limited data on the toxicity and occurrence of "emerging" mycotoxins. These mycotoxins are neither routinely determined nor legislatively regulated. Their presence has been reported in cereals from several countries [19–23]. In a recent EFSA report [24], an opinion on the presence of ENNs and BEA in food and feed was made, but the lack of relevant toxicity data did not a risk assessment. Currently, maximum levels for emerging *Fusarium* mycotoxins are not been regulated.

Maize is one of the most susceptible cereals to the presence of *Fusarium* molds. Infection of maize may lead to grain size and protein decreasing as well as harming germination. The final result is a decrease in yield and feed quality. Additionally, a consequential mycotoxin production is another highly problematic outcome of *Fusarium* infection.

In Serbia, arable land covers approximately 75.5% of utilized agricultural land. In the structure of sown arable land areas, cereals comprised 67.9%, industrial crops comprised 15.7%, vegetables comprised 2.6%, and fodder crops comprised 9.1% in 2016 [25]. However, cereals were grown on 1,763,575 ha in 2016, which is lower compared to 2015 (1,782,010 ha), and 2014 (1,819,188 ha). In 2016, maize was harvested from 1,010,097 ha, with a total production of 73,767,371 t. The average yield in 2016 was 7.3 t/ha, which was higher than in 2015 (5.4 t/ha), and slightly lower in comparison with 2014 (7.5 t/ha) [25]. When compared to 2016, the total production of maize decreased by 45.5% in 2017, while the average yield was only 4.0 t/ha [26]. In 2018, expected production of maize was 6,965,000 t, which was 73.3% higher than in 2017, with an average yield ofapproximately 7.6 t/ha [27]. High maize production in 2018 positioned Serbia among the top ten maize exporters [28] and among the top twenty maize producers in the world [29].

In considering these facts, the aim of this study was to determine the current state of the level of emerging *Fusarium* mycotoxins in Serbian maize. Additionally, an effort was made to relate the obtained results with the weather conditions recorded during the trial period.

2. Results

2.1. Occurrence of Emerging Toxins in Maize Samples

Maize samples collected during the 2016 harvest were analyzed and the results are shown in Table 1. MON and BEA had the highest presence among emerging mycotoxins (>80%), except in the West-Backa region (50%). Other emerging mycotoxins were not detected at all or were found in fewer samples (<20%). Overall, maize samples from the Middle-Banat region were the most contaminatedfor all investigated emerging mycotoxins. MON, BEA, and FUS were present in all regions. Mean levels of MON ranged from 189.97 µg/kg (West-Backa) to 920.10 µg/kg (Srem). BEA mean levels were between 6.82 μg/kg (West-Backa) and 34.79 μg/kg (Srem). FUS levels were the highest among all tested mycotoxins. They ranged from 328.50 μg/kg in South-Backa to 12,272.00 μg/kg in a sample from the West-Backa region. ENs were found in all regions except Srem, with the highest mean levels in samples originating from Middle-Banat.

Table 1. Occurrence of emerging toxins in maize samples collected in the Republic of Serbia in 2016.

	MON	BEA	EN A	EN A1	EN B	EN B1	FUS
South-Banat region							
Average ± SD (μg/kg)	237 ± 230	18.7 ± 30.3	-	0.59	7.55	4.86	328 ± 268
Range (μg/kg)	5.06–850	0.41–129	-	-	-	-	85.4–1121
Samples	21	21	21	21	21	21	21
Positive samples (%)	21 (100.0)	20 (95.2)	0	1 (4.8)	1 (4.8)	1 (4.8)	16 (76.2)
South-Backa region							
Average ± SD (μg/kg)	534 ± 410	13.7 ± 26.7	0.25 ± 0.19	0.24 ± 0.18	-	-	827 ± 1032
Range (μg/kg)	15.3–1450	0.10–111	0.12–0.47	0.13–0.44	-	-	91.3–4687
Samples	29	29	29	29	29	29	29
Positive samples (%)	26 (89.7)	26 (89.7)	3 (10.3)	3 (10.3)	0	0	22 (75.9)
Middle-Banat region							
Average ± SD (μg/kg)	576 ± 391	7.06 ± 14.4	8.78 ± 11.8	9.30 ± 15.7	0.80 ± 1.01	8.27 ± 11.4	1018 ± 396
Range (μg/kg)	7.18–1228	0.23–49.7	0.41–17.1	0.11–27.4	0.08–1.52	0.20–16.3	450–1738
Samples	12	12	12	12	12	12	12
Positive samples (%)	12 (100.0)	11 (91.7)	2 (16.7)	3 (25.0)	2 (16.7)	2 (16.7)	11 (91.7)
Srem region							
Average ± SD (μg/kg)	920 ± 1649	34.8 ± 67.8	-	-	-	-	1736 ± 2384
Range (μg/kg)	3.03–3856	0.27–136	-	-	-	-	312–4488
Samples	5	5	5	5	5	5	5
Positive samples (%)	5 (100.0)	4 (80.0)	0	0	0	0	3 (60.0)
West-Backa region							
Average ± SD (μg/kg)	190 ± 192	6.82 ± 9.90	0.49	0.53	-	0.22	12272
Range (μg/kg)	34.8–405	0.03–18.2	-	-	-	-	-
Samples	6	6	6	6	6	6	6
Positive samples (%)	3 (50.0)	3 (50.0)	1 (16.7)	1 (16.7)	0	1 (16.7)	1 (16.7)

The results of maize samples collected during 2017 are summarized in Table 2. ENs were not found in any of the four regions. MON, BEA, and FUS were found in all regions, except for FUS, which was not found in the sample from North-Backa. All the samples from the South-Banat region were contaminated with MON and BEA. The highest mean levels of MON (499.00 μg/kg) and FUS (3415.88 μg/kg) were recorded in the West-Backa region, while the highest mean level of BEA (12.26 μg/kg) was recorded in the South-Banat region.

Table 2. Occurrence of emerging toxins in maize samples collected in the Republic of Serbia in 2017.

	MON	BEA	EN A	EN A1	EN B	EN B1	FUS
South-Banat region							
Average ± SD (μg/kg)	404 ± 493	12.3 ± 18.9	-	-	-	-	353 ± 414
Range (μg/kg)	10.4–1803	0.22–67.4	-	-	-	-	63.1–1275
Samples	17	17	17	17	17	17	17
Positive samples (%)	17 (100.0)	17 (100.0)	0	0	0	0	10 (58.8)
South-Backa region							
Average ± SD (μg/kg)	179 ± 255	6.30 ± 16.4	-	-	-	-	468 ± 843
Range (μg/kg)	2.68–1071	0.04–75.9	-	-	-	-	45.5–3018
Samples	33	33	33	33	33	33	33
Positive samples (%)	20 (60.6)	21 (63.6)	0	0	0	0	12 (36.4)
West-Backa region							
Average ± SD (μg/kg)	499 ± 880	1.96 ± 3.88	-	-	-	-	3416 ± 9786
Range (μg/kg)	1.66–2999	0.06–13.4	-	-	-	-	40.6–29512
Samples	21	21	21	21	21	21	21
Positive samples (%)	12 (57.1)	11 (52.4)	0	0	0	0	9 (42.9)

Table 2. *Cont.*

	MON	BEA	EN A	EN A1	EN B	EN B1	FUS
North-Backa region							
Average ± SD (µg/kg)	221	18.3	-	-	-	-	-
Range (µg/kg)	-	-	-	-	-	-	-
Samples	1	1	1	1	1	1	1
Positive samples (%)	1 (100.0)	1 (100.0)	0	0	0	0	0

In 2018, MON was found in all three regions (South-Backa, North-Backa, and South-Banat), BEA and FUS were not found in samples from the South-Banat region and ENs were not present in any sample (Table 3). The mean levels of all tested mycotoxins were highest in the South-Backa region (MON 199.32 µg/kg, BEA 4.89 µg/kg and FUS 5793.79 µg/kg).

Table 3. Occurrence of emerging toxins in maize samples collected in the Republic of Serbia in 2018.

	MON	BEA	EN A	EN A1	EN B	EN B1	FUS
South-Banat region							
Average ± SD (µg/kg)	39.5 ± 44.6	-	-	-	-	-	-
Range (µg/kg)	5.19–89.9	-	-	-	-	-	-
Samples	6	6	6	17	6	6	6
Positive samples (%)	3 (50.0)	0	0	0	0	0	0
South-Backa region							
Average ± SD (µg/kg)	199 ± 238	4.89 ± 6.91	-	-	-	-	5794 ± 14,479
xRange (µg/kg)	5.80–857	0.15–21.5	-	-	-	-	63.2–38,610
Samples	34	34	34	34	34	34	34
Positive samples (%)	16 (47.1)	9 (26.5)	0	0	0	0	7 (20.6)
North-Backa region							
Average ± SD (µg/kg)	34.4 ± 36.9	3.33	-	-	-	-	72.4
Range (µg/kg)	55.82–88.7	-	-	-	-	-	-
Samples	5	5	5	5	5	5	5
Positive samples (%)	4 (80.0)	1 (20.0)	0	0	0	0	1 (20.0)

2.2. Climate Conditions

Reports from the Republic Hydrometeorological Service of Serbia [30] showed that the vegetation period of 2016 (April–September) in the territory of Serbia was warmer with somewhat higher precipitation than the long-term average. The deviation of mean daily temperatures during the vegetation period showed positive values (0.8 °C to 1.6 °C). The standardized precipitation index (SPI-3), determined for the summer period from 1 June to 31 August, showed normal humidity conditions for most of the territory of Vojvodina. However, in some parts of Vojvodina, moderate to extremely humid conditions were recorded (Figure 1B). Such conditions were registered in the Middle-Banat and West-Backa regions, and some parts of the South-Backa and South-Banat regions. In some production areas, strong winds and hail storms were recorded and certainly contributed to the damage of the grains and the occurrence of fungi on crops. The moisture in deeper soil layers in the middle of June was significantly reduced in Vojvodina as a result of a weaker inflow of precipitation in these areas.

If the weather conditions were observed in more detail, warm but unstable weather prevailed during the transition from May to June, and the agrometeorological conditions allowed the intensive development of maize. The trend of variable but warm weather continued in June. In the middle of the month, due to the influx of very hot air, the temperatures were considerably above the average for this period of the year. Maximum daily temperatures reached 36 °C on some days [30]. Thermal conditions were favorable for the intensive development of spring agricultural crops. By the end of the first decade of August, the weather was mostly dry and stable, but since the beginning of the second decade of August, the air temperatures moved around and were below average values. Maximum air temperatures were up to 28 °C, while the minimum morning temperatures were significantly below the average values for this time of the year [30]. Significant precipitation, mostly rain showers, was recorded on the territory of the entire country. During July and August, 2 to 3 times more rain was registered in the territory of Serbia compared to the average quantities.

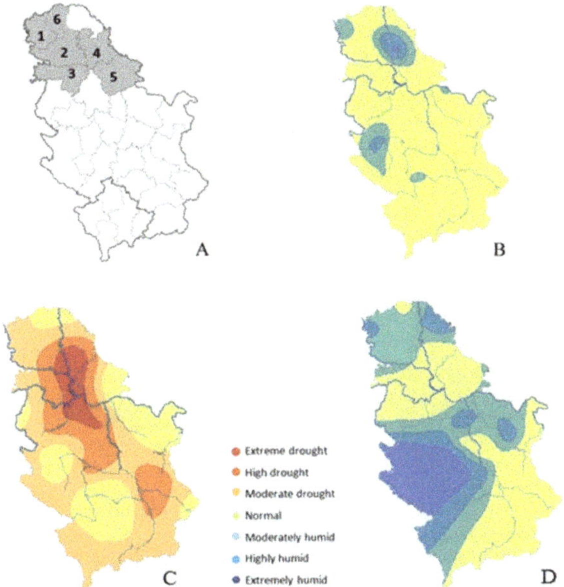

Figure 1. Regions of sample origin: **1**—West-Backa, **2**—South-Backa, **3**—Srem, **4**—Middle-Banat, **5**—South-Banat, **6**—North-Backa. (**A**) Humidity conditions in Serbia based on the Standardized Precipitation Index (SPI-3) determined for the summer period from 1 June to 31 August in 2016 (**B**) Reproduced from Agrometeorološki uslovi u proizvodnoj 2015/2016 godini, 2017, Republic Hydrometeorological Service of Serbia [30], 2017 (**C**) Reproduced from Agrometeorološki uslovi u proizvodnoj 2016/2017 godini, 2018, Republic Hydrometeorological Service of Serbia [31], and 2018 (**D**) Reproduced from Agrometeorološki uslovi u proizvodnoj 2017/2018, 2019, Republic Hydrometeorological Service of Serbia [32].

Detailed data on precipitation and temperature obtained from Metos®automatic weather stations (Metos®, Pessl Instruments, Weiz, Austria) in the observed regions in 2016 were compared to the multiannual average for 1981–2010 [33].Precipitation data (Figure 2) show that the Middle-Banat region, which overall was the most contaminated with emerging *Fusarium* toxins, and Srem, which showed the highest values of toxins, had precipitation values higher than the average in June, but they were noticeably lower in July. Average daily air temperature data (Figure 2) show that the temperature was around or slightly above the long-term average until August, when the temperature decreased. Average daily air temperatures were similar in all regions.

Reports from the Republic Hydrometeorological Service of Serbia [31] showed that the vegetation period of 2017 (April–September) was warmer and dryer than the multiannual average. The mean daily temperatures were 0.9–1.7 °C higher than the average, while precipitation was 20% lower than the average. SPI-3 showed extreme drought in the South-Banat region, while other observed regions were affected by high to moderate drought (Figure 1C).

The vegetation period started with unusually cold weather in April, but the weather conditions quickly normalized and became optimal for plant development during May [31]. During June, mean daily temperatures were higher than the average [31]. Maximum daily temperatures reached over 35 °C on some days, especially during the first and the last decade of the month. Hot weather continued during July. In most regions, precipitation was below the multiannual average, especially in the

South-Banat region, where it was 50% lower than the average. The temperatures at the beginning of August were extremely high (38–42 °C) and higher than the average during the whole month [31]. Most regions were affected with drought, except the South-Banat region, which had 50% more rainfall than the average.

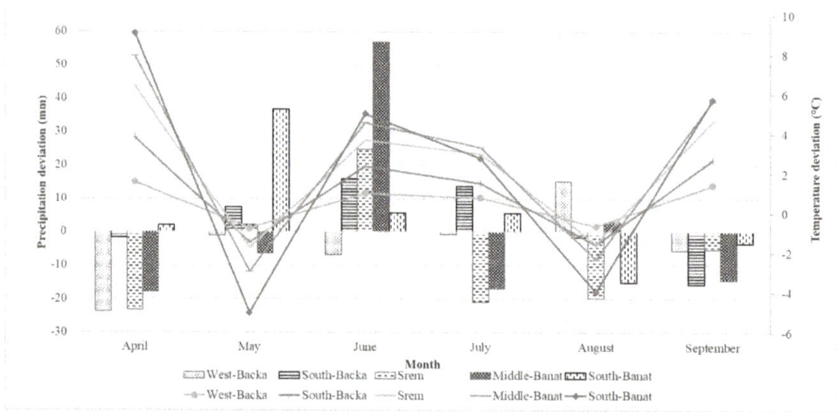

Figure 2. Deviation of total rainfall amount (columns) and average daily air temperature (lines) from the multiannual average (1981–2010) in 2016.

Precipitation and average daily air temperature data obtained from Metos®automatic weather stations (Metos®, Pessl Instruments, Weiz, Austria) in the observed regions during 2017 were compared to the multiannual average from 1981–2010 [33]. Precipitation data (Figure 3) showed that all regions had lower precipitation than the average during June. This trend continued through July for all regions except West-Backa, where high occurrences of MON and FUS were observed. The deviation of average daily air temperature (Figure 3) shows that temperature in April was lower than the multiannualaverage in all regions. During May, it was slightly above the multiannualaverage in all regions except South-Banat. In June the temperature was 1.4–2.4 °C higher than the average. Temperatures continued to be above the average until September.

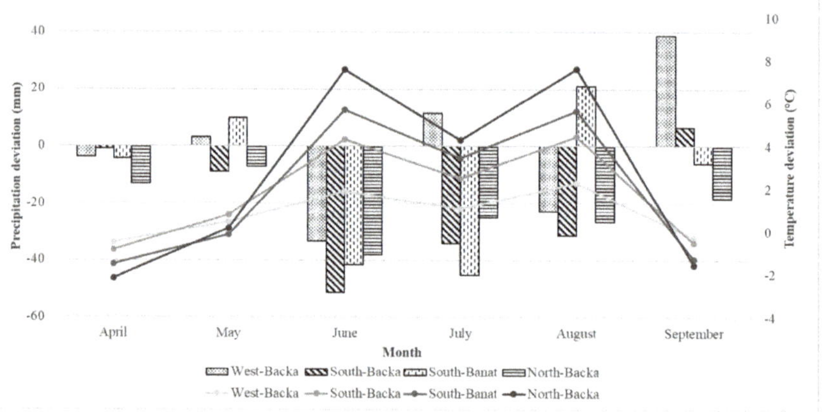

Figure 3. Deviation of total rainfall amount (columns) and average daily air temperature (lines) from multiannual average (1981–2010) in 2017.

According to reports from the Republic Hydrometeorological Service of Serbia [32], the vegetation period in 2018 (April–September) was 1.8–2.6 °C warmer than the multiannual average. SPI-3 showed normal to moderately humid weather conditions in all observed regions (Figure 1D).

The vegetation period started with unusually warm weather in April when the mean daily temperature was 4–5 °C above the multiannualaverage [32]. During May, weather conditions were optimal for plant growth. In the beginning of June, maximum daily temperatures were high (28–34 °C), but temperature decreased in the last decade of the month when maximum daily temperatures were in the range from 19 °C to 24 °C [32]. Frequent rain showers were recorded during June and continued through July. Warm weather with temperatures above multiannualaverage continued through August and September [32].

The deviation of precipitation and average daily air temperature data collected from Metos®automatic weather stations (Metos®, Pessl Instruments, Weiz, Austria) during 2018 from the multiannual average of 1981–2010 [33] was observed. Precipitation data (Figure 4) showed that precipitation was below average during the whole vegetation period in the North-Backa region. Precipitation was lower than the average in all three regions in April, May, August, and September. In June, precipitation was above average in the South-Backa and South-Banat regions. South-Backa was the only region where high precipitation continued in July.Average daily air temperature data (Figure 4) show that the temperature was above the long-term average in all regions during April and May. Temperature decreased during June and in July, reaching the multiannualaverage value in the South-Backa region, while the North-Backa and South-Banat regionswere below average. In August and September, the temperature increased again above the multiannual average in all observed regions.

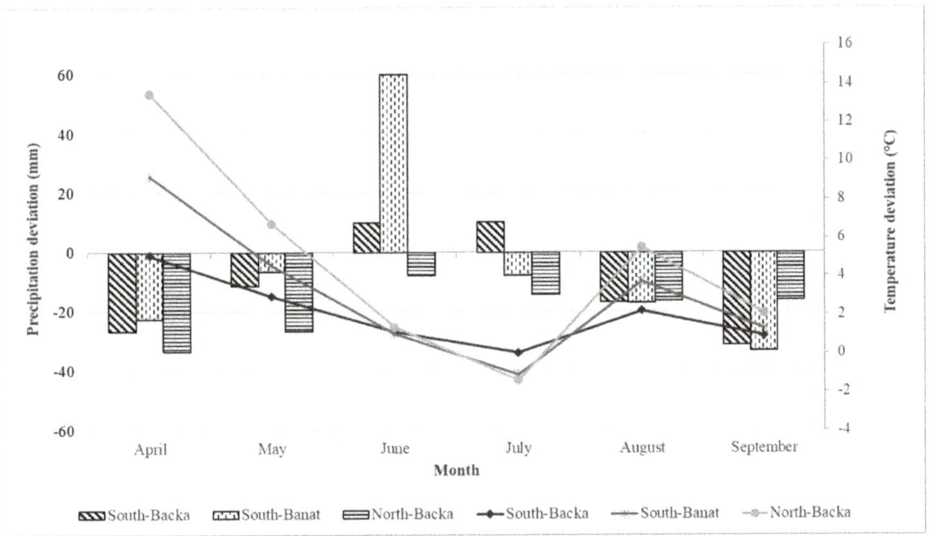

Figure 4. Deviation of total rainfall amount (columns) and average daily air temperature (lines) from the multiannual average (1981–2010) in 2018.

2.3. Statistical Analysis

The Kruskal–Wallis test (95% confidence level) found significant differences in MON levels deriving from different years and different regions. Additionally, MON levels significantly differed when compared in terms of average seasonal temperature and precipitation. For FUS, significant differences were found between years, average seasonal temperatures and precipitation, but not between regions. Statistically significant differences in BEA levels were found between years and regions. Furthermore, the Spearman correlation determined a slightly moderate negative linear correlation between average seasonal temperatures and mycotoxin contamination levels ($r = -0.41$ for MON, $r = -0.5$ for BEA, and $r = -0.45$ for FUS). For monthly weather data, only temperatures in May showed a moderate negative linear correlation ($r = -0.5$ for MON, r $= -0.58$ for BEA, and r $= -0.48$ for FUS), while precipitation in May showed a moderate positive linear correlation with contamination levels of observed mycotoxins ($r = 0.51$ for MON, $r = 0.59$ for BEA, and $r = 0.49$ for FUS). Stepwise regression showed (with 95% confidence level) that air temperatures from May to August had statistically significant differences in MON levels. For FUS, temperatures and precipitation in any month, as well as average values for the whole growing season, did not show statistically significant differences among contamination levels. May temperatures had a statistically significant difference in BEA levels. The adjusted R-squared values were between 0.101 and 0.2278 and showed that these models cannot be used as prediction models; however, they gave insight about statistically significant variables that influence mycotoxin contamination.

3. Discussion

Sutton [34] explained that for maize, *Fusarium* infection of the ear most frequently takes place through the tip of the ear when the fungi penetrate through the silk during maize flowering. Very humid weather during the period from silking to ripening enables ear contamination [35]. The ear is the most susceptible to contamination at the beginning of silking, while the susceptibility decreases with silk aging [36,37]. The silking period in the climatic region of Serbia takes place during July and the first half of August.

According to the Republic Hydrometeorological Service of Serbia [30], most of the critical period for *Fusarium* infection of maize (July–August) in 2016 was characterized as dry and stable weather. However, since the beginning of the second decade of August, air temperatures ranged around and below average values [30]. Maximum air temperatures were up to a maximum of 28 °C, while the minimum morning temperatures were significantly below the average values for this time of the year. However, moderate to extremely humid conditions occurred during the summer of 2016 in the Middle-Banat and West-Backa regions and some parts of the South-Backa and South-Banat regions, which may have led to *Fusarium* fungi growth and the consequent production of mycotoxins. Moreover, cool, cloudy and humid weather during July and August did not favor agricultural crops and such conditions probably caused plant stress and higher susceptibility to *Fusarium* infection.

In 2017, precipitation was lower than the multiannual average during summer months in all regions except South-Banat, where high levels of BEA were recorded. High precipitation in these two regions during silking (July and beginning of August), when the maize is the most susceptible to *Fusarium* infections were favorable for fungal development, may have led to high levels of emerging toxins in samples from these regions.

High precipitation during June 2018 in the South-Backa and South-Banat regions and during July 2018 in South-Backa resulted in South-Backa having the highest mean levels of MON, BEA, and FUS. Warm weather during July, together with high humidity in the South-Backa region enabled *Fusarium* infection of the ears, which may be related to the high mycotoxin contamination of samples from this region.

Emerging fusariotoxins were mostly investigated in Mediterranean countries. Juan et al. [22] analyzed 93 samples of organic cereals and organic cereal products from several local markets in Italy for the presence of different mycotoxins, including BEA, EN A, EN A1, EN B, EN B1. The authors found that levels of some emerging *Fusarium* mycotoxins ranged as follows: BEA 6.7–41 µg/kg, EN A 7.2–29.8 µg/kg, EN A1 5.3–64.3 µg/kg, EN B 5.5–102 µg/kg and EN B1 5.5–33.1 µg/kg. Among the commodities, the occurrence was the highest in wheat samples. Serrano et al. [21], investigated fusariotoxins' occurrence in the Mediterranean area. They found that BEA was present in 2 of 14 maize samples and in 1 of 22 maize-based products. Obtained levels were 2.1 and 73.9 µg/kg for maize and 5.2 µg/kg for maize-based products, respectively. A high divergence among detected BEA levels was also found in this study (0.03–136 µg/kg). Later, Serrano et al. compared levels of emerging fusariotoxins between organic and conventional pasta in Spain [38]. They found that organic pasta was more contaminated than the conventional type. Contamination levels were 0.10–20.96 µg/kg for BEA, and 0.05–8.02 µg/kg for FUS, while ENs levels were 0.25–979.56 µg/kg. Remarkably high levels of emerging mycotoxins in raw cereals were found in Morocco by Zinedine et al. [39]. EN A1 was predominant among ENs with a presence in 39% of samples and levels ranging from 14 to 445 mg/kg. BEA was found in 26.5% of samples, with levels ranging from 1 to 59 mg/kg, while FUS was present in 7.8% of samples (levels from 0.6 to 2 mg/kg). Regarding maize samples, 42% contained ENs with mean levels of 207 mg/kg (EN A1), 54 mg/kg (EN B), 8 mg/kg (EN B1), while EN A was not detected. A similar situation was observed in cereals from the Spanish market [40]. The authors reported very a high presence of ENs (73.4%), wherein EN A1 was the most frequent with the highest levels (33.36–814.42 mg/kg). BEA was found in 32.8% of samples in the range of 0.51–11.78 mg/kg, and FUS levels were between 1.01–6.63 mg/kg with the presence in 7.8% of samples. In maize samples, the presence of ENs was 89%, BEA was found in 21% and FUS was in only one sample; on the other hand, ENs were found only during 2016 in this study, but not in any samples from 2017 and 2018, while BEA and FUS were detected every year. The highest mean level in maize was obtained for EN A1 of 813.01 mg/kg, while in this survey the highest mean level of EN A1 was only 9.30 µg/kg. In another study in Morocco on maize-based breakfast cereals, Mahnine et al. [41] obtained mean levels of 113 mg/kg for EN A1 and 20.1 mg/kg for EN B1, respectively, while EN A, EN B, FUS, and BEA were below LOQ in all samples. Tunisian cereals were highly contaminated as well. Maize-based cereals only contained EN A1 and ENB1 with mean levels of 113 mg/kg and 20.1 mg/kg. Oueslati et al. [20] obtained the presence of ENs in 96% of samples, where once more EN A1 was predominant (92.1%). Mean values were the highest in the case of EN A1 (up to 480 mg/kg). Only 3 samples of maize were analyzed. Two samples were positive, one with EN A1 (29.6 mg/kg) and another contained EN B1 (17.0 mg/kg). Notably, none of these authors analyzed results along with the weather conditions. Emerging fusariotoxins were also studied in rice. In Morocco, considerable contamination with ENs and BEA was revealed, but not with FUS [19]. In rice samples from Iran, a significant presence was found only in the case of BEA (40%), but in very low amounts [23].

Emerging fusariotoxins were also studied in some non-Mediterranean European countries. Goertz et al. [42] investigated the contamination of different maize hybrids in Germany during 2006 and 2007. BEA was found in 52% of samples from 2006 and in 33% of samples from 2007. Mean levels were 390 µg/kg and 240 µg/kg, respectively. MON was detected in 45% and 43% of samples, respectively, with mean values of 280 µg/kg and 110 µg/kg, respectively. Among EN, only EN B was investigated. Although its presence was relatively high (41% and 30%, respectively) the levels were the lowest (mean of 70 µg/kg and 160 µg/kg, respectively) among investigated emerging fusariotoxins. The authors explained that moderate temperatures and frequent precipitation recorded during early growth stages in 2007 were favorable for *Fusarium* growth. This is in accordance with the results of correlation analysis in this study, which showed a moderately negative correlation between May temperatures and toxin contamination, together with a moderately positive correlation between May precipitation and toxin contamination. However, a higher mycotoxin presence and higher contamination levels in Germany were found in samples from 2006, which they associated with maize exposure to drought stress in July and September 2006. In Norway, Uhlig et al. [43] investigated MON occurrence in Norwegian grain (oats, barley, and wheat) during a three-year period (2000–2002). MON was found in 46% of samples and the obtained levels were between 43 and 950 µg/kg. The authors noted that the highest prevalence of MON was found in the 2002 season (67%), along with the highest concentration. In Poland, Chelkowski et al. [44] detected ENs and BEA in 18 out of 27 maize samples (levels of 0.8–46.0 mg/kg). Unfortunately, in both studies, the weather conditions were not discussed.

To the best of our knowledge, studies on emerging *Fusarium* toxins in Serbia have not been done to date. On the other hand, some studies have occurred in surrounding countries. In Romania, Stanciu et al. [45] found that ENs were the most frequent (73%) mycotoxins in both wheat and wheat flour, while EN B was detected the most (71%). The highest observed concentration was 407 µg/kg in wheat samples. Mean values were 19 µg/kg in wheat flour and 128 µg/kg in wheat. In neighboring Croatia, Jurjević et al. [46] investigated BEA presence in 209 maize samples originating from the 1996 and 1997 growing seasons. The authors found that 17.4% of samples from 1996 contained BEA at the mean level of 393 µg/kg and maximum concentration of 1864 µg/kg. In samples from 1994, only one of 104 samples contained BEA.

Based on the obtained results and available published data, the results from this study are in accordance with those found in Croatia, Italy, and Germany, while results from Poland, Spain, Morocco, and Tunisia are one order of magnitude higher. Unfortunately, studies from Romania, Norway, and Iran did not include maize or maize-based products and therefore a valid comparison cannot be made.

4. Conclusions

The main source of emerging *Fusarium* mycotoxins are cereals that are used in food and feed production, and they may thus pose a potential risk for human and animal health. Since environmental conditions in Serbia are favorable for the occurrence of mycotoxigenic fungi, monitoring of "traditional" but also "emerging" *Fusarium* toxins is essential for producing safe food and feed. The results indicated that most attention should be paid to fusaproliferin (FUS) and moniliformin (MON). Additionally, monitoring studies for emerging *Fusarium* mycotoxins are necessary for legislative purposes, because in the near future appropriate maximum contamination levels should be set for several mycotoxins by relevant authorities [38].

5. Materials and Methods

5.1. Samples

In total, 190 representative samples from commercial fields in Serbia were analyzed. Samples were collected during harvest in the northern Serbian province of Vojvodina, which is the country's most important agricultural area, over three years: 73 samples from 28 localities in 2016, 72 samples from 12 localities in 2017 and 45 samples from 13 localities in 2018. Localities were clustered based on their administrative area into 6 regions: West-Backa, South-Backa, Srem, Middle-Banat, South-Banat, and North-Backa (Figure 1A).

Each sample was transported to the laboratory immediately after sampling and stored in a freezer at −20 °C until analysis. Prior to analysis, the samples were allowed to reach room temperature. All samples were milled on a laboratory mill so that >93% passed through a sieve with a pore diameter of 0.8 mm and a portion was taken for analysis.

5.2. Extraction and Mycotoxin Analysis in Maize Samples

Five grams of each milled sample were extracted using a 20 mL extraction solvent (acetonitrile–water–acetic acid (VWR, Vienna, Austria), 79:20:1, *v/v/v*) followed by a 1 + 1 dilution using acetonitrile–water–acetic acid (VWR, Vienna, Austria) (20:79:1, *v/v/v*) and a direct injection of 5 µL diluted extract.

Liquid chromatography-tandem mass spectrometry (LC-MS/MS) screening of target fungal metabolites was performed at theInstitute of Bioanalytics and Agro-Metabolomics, Department of Agrobiotechnology (IFA-Tulln), University of Natural Resources and Life Sciences, Vienna, with a QTrap 5500 LC-MS/MS System (Applied Biosystems, Foster City, CA, USA) equipped with a TurboIon Spray electrospray ionization (ESI) source and a 1290 Series HPLC System (Agilent, Waldbronn, Germany). Chromatographic separation was performed at 25 °C on a Gemini® C_{18}-column, 150 × 4.6 mm i.d., 5 µm particle size, equipped with a C_{18} 4 × 3 mm i.d. security guard cartridge (all from Phenomenex, Torrance, CA, USA). The chromatographic method, chromatographic and mass spectrometric parameters, as well as the method validation data, are described by Malachova et al. [47]. Electrospray ionization-tandem mass spectrometry (ESI-MS/MS) was performed in the time-scheduled multiple reaction monitoring (MRM) mode both in positive and negative polarities in two separate chromatographic runs per sample by scanning two fragmentation reactions per analyte. The MRM detection window of each analyte was set to its expected retention time of ±27 sec and ±48 sec in the positive and the negative mode, respectively. Confirmation of a positive analyte identification was obtained by the acquisition of two MRMs per analyte (excepting moniliformin, which exhibits only one fragment ion), which yielded 4.0 identification points according to commission decision 2002/657/EC. In addition, the LC retention time and the intensity ratio of the two MRM transitionswas in accordance with the related values of an authentic standard within 0.1 min and 30% rel., respectively.

Quantification was based on an external calibration using a serial dilution of a multianalyte stock solution, and results were corrected for apparent recoveries. The accuracy of the method is verified on a continuous basis by regular participation in proficiency testing schemes [47,48] organized by BIPEA (Gennevilliers, France). Based on the submitted results, a general expanded measurement uncertainty of 50% has been determined [49]. In the case of the 175 results already submitted for maize and maize-based feed, 168 results were in the satisfactory range (z-score between −2 and 2).

5.3. Statistical Analysis

Statistical analysis (Supplementary Materials) was performed using the computing environment *R* (*R* Core Team, Vienna, Austria) [50] on the data from regions where samples were collected during all three years of research (South-Backa and South-Banat). The Shapiro–Wilk normality test was used to check the distribution of the data. Since the data were not normally distributed, nonparametric tests were used for further analysis. The Kruskal–Wallis test was used to check whether the mean ranks of the mycotoxin levels were the same in all groups. Spearman's correlation was used to determine the correlation between climate conditions and mycotoxin contamination levels. Furthermore, Stepwise regression with backward steps was used to obtain the optimal model and significant months in terms of temperature and precipitation values that influence mycotoxin levels.

Supplementary Materials: The following are available online at http://www.mdpi.com/2072-6651/11/6/357/s1, Figure S1. Spearman Correlation Results (*r*).

Author Contributions: Conceptualization, I.J. and F.B.; Data curation, T.D., S.K., Z.S., and A.S.; Formal analysis, T.D., S.K., and D.G.; Funding acquisition, I.J., R.K., and F.B.; Investigation, R.K., M.S., Z.S., D.G., and A.S.; Methodology, R.K., M.S., and D.G.; Project administration, I.J. and F.B.; Resources, I.J. and F.B.; Software, T.D. and S.K.; Supervision, I.J., R.K., and F.B.; Validation, I.J., R.K. and M.S.; Visualization, T.D. and S.K.; Writing—originaldraft, T.D. and S.K.; Writing—review and editing, I.J., R.K., F.B., and Z.S.

Funding: This research was funded by MyToolBox (EU's Horizon 2020 research and innovation programme, agreement No 678012), research project of the Ministry of Education, Science and Technological Development, Republic of Serbia (project III 46005) and Provincial Secretariat for Higher Education and Scientific Research of the Autonomous Province of Vojvodina through the project "Application of novel and conventional processes for removal of most common contaminants, mycotoxins and salmonella, in order to produce safe animal feed inthe territory of AP Vojvodina", Project No. 114-451-2505/2016-01.

Conflicts of Interest: The authors declare no conflict of interest.

References

1. Jestoi, M. Emerging Fusarium mycotoxins fusaproliferin, beauvericin, enniatins, and moniliformin: A review. *Crit. Rev. FoodSci. Nutr.* **2008**, *48*, 21–49. [CrossRef] [PubMed]

2. Kumar, V.; Basu, M.S.; Rajendran, T.P. Mycotoxin research and mycoflora in some commercially important agricultural commodities. *Crop Prot.* **2008**, *27*, 891–905. [CrossRef]

3. Kovalsky, P.; Kos, G.; Nährer, K.; Schwab, C.; Jenkins, T.; Schatzmayr, G.; Sulyok, M.; Krska, R. Co-occurrence of regulated, masked and emerging mycotoxins and secondary metabolites in finished feed and maize—An extensive survey. *Toxins* **2016**, *8*, 363. [CrossRef] [PubMed]

4. Burmeister, H.R.; Ciegler, A.; Vesonder, R.F. Moniliformin, a metabolite of *Fusarium moniliforme* NRRL 6322: Purification and toxicity. *Appl. Environ. Microbiol.* **1979**, *37*, 11–13. [PubMed]

5. Cole, R.J.; Kirksey, J.W.; Cutler, H.G.; Doupnik, B.L.; Peckham, J.C. Toxin from *Fusarium moniliforme*: Effects on plants and animals. *Science* **1973**, *179*, 1324–1326. [CrossRef] [PubMed]

6. Calo, L.; Fornelli, F.; Ramiers, R.; Nenna, S.; Tursi, A.; Caiaffa, M.F.; Macchia, L. Cytotoxic effects of the mycotoxin beauvericin to human cell lines of myeloid origin. *Pharmacol. Res.* **2004**, *49*, 73–77. [CrossRef] [PubMed]

7. Ivanova, L.; Skjerve, E.; Eriksen, G.S.; Uhlig, S. Cytotoxicity of enniatins A, A1, B, B1, B2 and B3 from *Fusarium avenaceum*. *Toxicon* **2006**, *47*, 868–876. [CrossRef] [PubMed]

8. Jow, G.M.; Chou, C.J.; Chen, B.F.; Tsai, J.H. Beauvericin induces cytotoxic effects in human acute lymphoblastic leukemia cells through cytochrome c release, caspase 3 activation: The causative role of calcium. *Cancer Lett.* **2004**, *216*, 165–173. [CrossRef]

9. Ferrer, E.; Juan-García, A.; Font, G.; Ruiz, M.J. Reactive oxygen species induced by beauvericin, patulin and zearalenone in CHO-K1 cells. *Toxicol. In Vitro* **2009**, *23*, 1504–1509. [CrossRef]

10. Ritieni, A.; Fogliano, V.; Randazzo, G.; Scarallo, A.; Logrieco, A.; Moretti, A.; Manndina, L.; Bottalico, A. Isolation and characterization of fusaproliferin, a new toxic metabolite from *Fusarium proliferatum*. *Nat. Toxins* **1995**, *3*, 17–20. [CrossRef]

11. Tomoda, H.; Huang, X.H.; Nishida, J.; Cao, H.; Nagao, R.; Okuda, S.; Tanaka, H.; Omura, S.; Arai, H.; Inoue, K. Inhibition of acyl-CoA: Cholesterol acyltransferase activity by cyclodepsipeptide antibiotics. *J. Antibiot.* **1992**, *45*, 1626–1632. [CrossRef] [PubMed]

12. Macchia, L.; Di Paola, R.; Fornelli, F.; Nenna, S.; Moretti, A.; Napoletano, R.; Logrieco, A.; Caiaffa, M.F.; Bottalico, A. Cytotoxicity of beauvericin to mammalian cells. In *Abstracts of the International Seminaron Fusarium: Mycotoxins, Taxonomyand Pathogenicity, Martina Franca, Italy, 9–13 May 1995*; Stampasud: Mottola, Italy, 1995; pp. 72–73.

13. Fornelli, F.; Minervini, F.; Logrieco, A. Cytotoxicity of fungal metabolites to lepidopteran (*Spodopterafrugiperda*) cell line (SF-9). *J. Invertebr. Pathol.* **2004**, *85*, 74–79. [CrossRef] [PubMed]

14. Ojcius, D.M.; Zychlinsky, A.; Zheng, L.M.; Young, J.D.E. Ionophore induced apoptosis: Role of DNA fragmentation and calcium fluxes. *Exp. Cell Res.* **1991**, *197*, 43–49. [CrossRef]

15. Randazzo, G.; Fogliano, V.; Ritieni, A.; Mannina, L.; Rossi, E.; Scarallo, A.; Segre, A.L. Proliferin, a new sesterterpene from *Fusarium proliferatum*. *Tetrahedron* **1993**, *49*, 10883–10896. [CrossRef]

16. Logrieco, A.; Moretti, A.; Fornelli, F.; Fogliano, V.; Ritieni, A.; Caiaffa, M.F.; Randazzo, G.; Bottalico, A.; Macchia, L. Fusaproliferin production by *Fusarium subglutinans* and its toxicity to *Artemia salina*, SF-9 insect cells, and IARC/LCL 171 human B lymphocytes. *Appl. Environ Microbiol.* **1996**, *62*, 3378–3384. [PubMed]

17. Ritieni, A.; Monti, S.M.; Randazzo, G.; Logrieco, A.; Moretti, A.; Peluso, G.; Ferracane, R.; Fogliano, V. Teratogenic effects of fusaproliferin on chicken. *J. Agric. Food Chem.* **1997**, *45*, 3039–3043. [CrossRef]

18. Kriek, N.P.J.; Marasas, W.F.O.; Steyn, P.S.; Van Rensburg, S.J.; Steyn, M. Toxicity of a moniliformin-producing strain of *Fusarium moniliforme* var. subglutinans isolated from maize. *Food Cosmet. Toxicol.* **1977**, *15*, 579–587. [CrossRef]

19. Sifou, A.; Meca, G.; Serrano, A.B.; Mahnine, N.; El Abidi, A.; Mañes, J.; El Azzouzi, M.; Zinedine, A. First report on the presence of emerging Fusarium mycotoxins enniatins (A, A1, B, B1), beauvericin and fusaproliferin in rice on the Moroccan retail markets. *Food Control* **2011**, *22*, 1826–1830. [CrossRef]

20. Oueslati, S.; Meca, G.; Mliki, A.; Ghorbel, A.; Mañes, J. Determination of Fusarium mycotoxins enniatins, beauvericin and fusaproliferin in cereals and derived products from Tunisia. *Food Control* **2011**, *22*, 1373–1377. [CrossRef]

21. Serrano, A.; Font, G.; Ruiz, M.; Ferrer, E. Co-occurrence and risk assessment of mycotoxins in food and diet from Mediterranean area. *Food Chem.* **2012**, *135*, 423–429. [CrossRef]

22. Juan, C.; Ritieni, A.; Mañes, J. Occurrence of Fusarium mycotoxins in Italian cereal and cereal products from organic farming. *Food Chem.* **2013**, *141*, 1747–1755. [CrossRef] [PubMed]

23. Nazari, F.; Sulyok, M.; Kobarfard, F.; Yazdanpanah, H.; Krska, R. Evaluation of emerging Fusarium mycotoxins beauvericin, enniatins, fusaproliferin and moniliformin in domestic rice in Iran. *Iran. J. Pharm. Res.* **2015**, *14*, 505–512. [PubMed]

24. EFSA CONTAM Panel (EFSA Panel on Contaminants in the Food Chain). Scientific Opinion on the risks to human and animal health related to the presence of beauvericin and enniatins in food and feed. *EFSA J.* **2014**, *12*, 3802. [CrossRef]

25. Statistical Office of the Republic of Serbia. Statistical Yearbook of Serbia 2017. Available online: http://publikacije.stat.gov.rs/G2017/Pdf/G20172022.pdf (accessed on 6 September 2018).

26. Statistical Office of the Republic of Serbia. Statistical Yearbook of Serbia 2018. Available online: http://publikacije.stat.gov.rs/G2018/Pdf/G20182051.pdf (accessed on 12 February 2019).

27. Statistical Office of the Republic of Serbia. Production of Wheat and Early Fruit and the Expected Yield of Late Crops, Fruit and Grape 2018. Available online: http://publikacije.stat.gov.rs/G2018/Pdf/G20181261.pdf (accessed on 26 September 2018).

28. Index Mundi: Corn Exports by Country 2018. Available online: http://www.indexmundi.com/agriculture/?commodity=corn&graph=exports (accessed on 8 March 2019).

29. Index Mundi: Corn Production by Country 2018. Available online: http://www.indexmundi.com/agriculture/?commodity=corn&graph=production (accessed on 8 March 2019).

30. Republic Hydrometeorological Service of Serbia. Agrometeorološki uslovi u proizvodnoj 2015/2016 godini na teritoriji Republike Srbije. Available online: http://www.hidmet.gov.rs/podaci/agro/ciril/AGROveg2016.pdf (accessed on 23 September 2017).

31. Republic Hydrometeorological Service of Serbia. Agrometeorološki uslovi u proizvodnoj 2016/2017 godini na teritoriji Republike Srbije. Available online: http://www.hidmet.gov.rs/podaci/agro/ciril/AGROveg2017.pdf (accessed on 1 November 2018).

32. Republic Hydrometeorological Service of Serbia. Agrometeorološki uslovi u proizvodnoj 2017/2018 godini na teritoriji Republike Srbije. Available online: http://www.hidmet.gov.rs/podaci/agro/godina.pdf (accessed on 20 February 2019).

33. Republic Hydrometeorological Service of Serbia. Multiannual Average of Meteorological Parameters (1981–2010). Available online: http://www.hidmet.gov.rs/ciril/meteorologija/klimatologija_srednjaci.php (accessed on 20 February 2019).

34. Sutton, J.C. Epidemiology of wheat head blight and maize ear rot caused by *Fusarium graminearum*. *Can. J. Plant Pathol.* **1982**, *4*, 195–209. [CrossRef]

35. Vigier, B.; Reid, L.M.; Seifert, K.A.; Stewart, D.W.; Hamilton, R.I. Distribution and prediction of *Fusarium* species associated with maize ear rot in Ontario. *Can. J. Plant Pathol.* **1997**, *19*, 60–65. [CrossRef]

36. Reid, L.M.; Bolton, A.T.; Hamilton, R.I.; Woldemariam, T.; Mather, D.E. Effect of silk age on resistance of maize to *Fusarium graminearum*. *Can. J. Plant Pathol.* **1992**, *14*, 293–298. [CrossRef]

37. Reid, L.M.; Hamilton, R.I. Effect of inoculation position, timing, macrocondial concentration, and irrigation on resistance of maize to *Fusarium graminearum* infection through kernels. *Can. J. Plant Pathol.* **1996**, *18*, 279–285. [CrossRef]

38. Serrano, A.B.; Font, G.; Mañes, J.; Ferrer, E. Emerging Fusarium mycotoxins in organic and conventional pasta collected in Spain. *Food Chem. Toxicol.* **2013**, *51*, 259–266. [CrossRef] [PubMed]

39. Zinedine, A.; Meca, G.; Mañes, J.; Font, G. Further data on the occurrence of Fusarium emerging mycotoxins enniatins (A, A1, B, B1), fusaproliferin and beauvericin in raw cereals commercialized in Morocco. *FoodControl* **2011**, *22*, 1–5. [CrossRef]

40. Meca, G.; Zinedine, A.; Blesa, J.; Font, G.; Mañes, J. Further data on the presence of Fusarium emerging mycotoxins enniatins, fusaproliferin and beauvericin in cereals available on the Spanish markets. *Food Chem. Toxicol.* **2010**, *48*, 1412–1416. [CrossRef] [PubMed]

41. Mahnine, N.; Meca, G.; Elabidi, A.; Fekhaoui, M.; Saoiabi, A.; Font, G.; Mañes, J.; Zinedine, A. Further data on the levels of emerging Fusarium mycotoxins enniatins (A, A1, B, B1), beauvericin and fusaproliferin in breakfast and infant cereals from Morocco. *Food Chem.* **2011**, *124*, 481–485. [CrossRef]

42. Goertz, A.; Zuehlke, S.; Spiteller, M.; Steiner, U.; Dehne, H.W.; Waalwijk, C.; de Vries, I.; Oerke, E.C. Fusarium species and mycotoxin profiles on commercial maize hybrids in Germany. *Eur. J. Plant Pathol.* **2010**, *128*, 101–111. [CrossRef]

43. Uhlig, S.; Torp, M.; Jarp, J.; Parich, A.; Gutleb §, A.C.; Krska, R. Moniliformin in Norwegian grain. *Food Addit. Contam.* **2004**, *21*, 598–606. [CrossRef] [PubMed]

44. Chelkowski, J.; Ritieni, A.; Wisniewska, H.; Mulé, G.; Logrieco, A. Occurrence of toxic hexadepsipeptides in preharvest maize ear rot infected by *Fusarium poae* in Poland. *J. Phytopathol.* **2007**, *155*, 8–12. [CrossRef]

45. Stanciu, O.; Juan, C.; Miere, D.; Loghin, F.; Mañes, J. Occurrence and co-occurrence of Fusarium mycotoxins in wheat grains and wheat flour from Romania. *Food Control* **2017**, *73*, 147–155. [CrossRef]

46. Jurjević, Ž.; Solfrizzo, M.; Cvjetković, B.; De Girolamo, A.; Visconti, A. Occurrence of beauvericin in corn from Croatia. *Food Technol. Biotech.* **2002**, *40*, 91–94.

47. Malachova, A.; Sulyok, M.; Beltran, E.; Berthiller, F.; Krska, R. Optimization and validation of a quantitative liquid chromatography—Tandem mass spectrometric method covering 295 bacterial and fungal metabolites including all regulated mycotoxins in four model food matrices. *J. Chromatogr. A* **2014**, *1362*, 145–156. [CrossRef] [PubMed]
48. Malachova, A.; Sulyok, M.; Beltrán, E.; Berthiller, F.; Krska, R. Multi-toxin determination in food - the power of "Dilute and Shoot" approaches in LC-MS-MS. *LCGC Eur.* **2015**, *28*, 542–555.
49. Stadler, D.; Sulyok, M.; Schuhmacher, R.; Berthiller, F.; Krska, R. The contribution to lot-to-lot variation to the measurement uncertainty of an LC-MS-based multi-mycotoxin assay. *Anal. Bioanal. Chem.* **2018**, *410*, 4409–4418. [CrossRef]
50. R Core Team. *R: A Language and Environment for Statistical Computing*; R Foundation for Statistical Computing: Vienna, Austria, 2013; ISBN 3-900051-07-0. Available online: http://www.R-project.org/ (accessed on 30 April 2019).

Article

Occurrence of the Ochratoxin A Degradation Product 2′R-Ochratoxin A in Coffee and Other Food: An Update

Franziska Sueck [1], Vanessa Hemp [1], Jonas Specht [1], Olga Torres [2,3], Benedikt Cramer [1,*] and Hans-Ulrich Humpf [1,*]

[1] Institute of Food Chemistry, Westfälische Wilhelms-Universität Münster, Corrensstraße 45, 48149 Münster, Germany; f_suec01@uni-muenster.de (F.S.); vhemp@web.de (V.H.); jonasspecht@uni-muenster.de (J.S.)
[2] Laboratorio Diagnostico Molecular S.A, Guatemala City, Guatemala; otorres@dxmolecular.com
[3] Centro de Investigaciones en Nutrición y Salud, Guatemala City, Guatemala
[*] Correspondence: cramerb@uni-muenster.de (B.C.); humpf@wwu.de (H.-U.H.)

Received: 18 April 2019; Accepted: 3 June 2019; Published: 8 June 2019

Abstract: Food raw materials can contain the mycotoxin ochratoxin A (OTA). Thermal processing of these materials may result in decreased OTA levels but also in the formation of the thermal isomerization product 2′R-ochratoxin A (2′R-OTA). So far, only 2′R-OTA levels reported from 15 coffee samples in 2008 are known, which is little when compared to the importance of coffee as a food and trading good. Herein, we present results from a set of model experiments studying the effect of temperatures between 120 °C and 270 °C on the isomerization of OTA to 2′R-OTA. It is shown that isomerization of OTA starts at temperatures as low as 120 °C. At 210 °C and above, the formation of 25% 2′R-OTA is observed in less than one minute. Furthermore, 51 coffee samples from France, Germany, and Guatemala were analyzed by HPLC-MS/MS for the presence of OTA and 2′R-OTA. OTA was quantified in 96% of the samples, while 2′R-OTA was quantifiable in 35% of the samples. The highest OTA and 2′R-OTA levels of 28.4 µg/kg and 3.9 µg/kg, respectively, were detected in coffee from Guatemala. The OTA:2′R-OTA ratio in the samples ranged between 2.5:1 and 10:1 and was on average 5.5:1. Besides coffee, 2′R-OTA was also for the first time detected in a bread sample and malt coffee powder.

Keywords: Ochratoxin A; 2′R-ochratoxin A; 14(R)-ochratoxin A; coffee; degradation; processing; roasting; modified mycotoxins; masked mycotoxins

Key Contribution: 2′R-OTA occurs in most of the tested coffee samples in levels up to 3.9 µg/kg. Other food might also contain significant amounts of 2′R-OTA as the isomerization of OTA to 2′R-OTA starts at 120 °C.

---

1. Introduction

The mycotoxin ochratoxin A (OTA, Figure 1) can be found in a broad spectrum of food raw materials infested with fungi of the genera *Aspergillus* and *Penicillium* as well as in food products derived from these commodities [1]. Considering exposure within the European Union, cereals and cereal products, such as pasta, bread, and beer, are the most relevant OTA sources due to the high consumption rates of these food items. In Germany, for instance, these products are on average responsible for about 67% of the OTA intake in adults. Besides this, approximately 12% of the total OTA exposure occurs from coffee drinking and a further 6% from cocoa, 6% from meat, and 5% from wine [1].

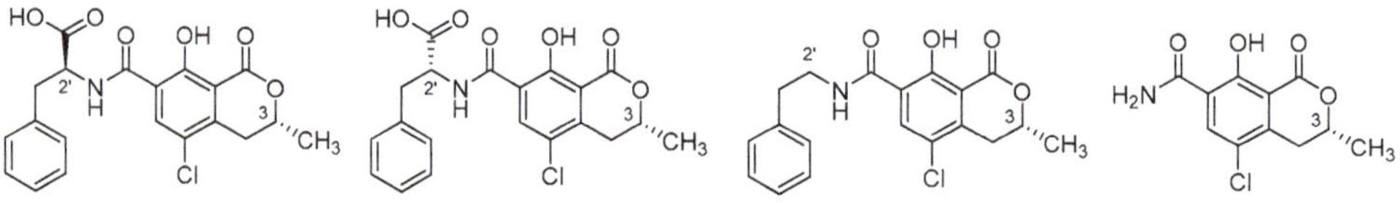

Figure 1. Structures of ochratoxin A (OTA), its thermal isomerization product 2′R-ochratoxin A (2′R-OTA), and the degradation products 2′-decarboxy-ochratoxin A (DC-OTA) and ochratoxin α-amide (OTamide).

Most food items undergo food processing, such as milling, baking, roasting, frying, or fermentation. All of these steps can have an impact on the mycotoxin burden of the product, for instance by physical removal or chemical modification [2,3]. To what extent mycotoxins are modified during these processing steps depends on the parameters but also on the chemical structure of the compound. Chemical modifications can result in the formation of mycotoxin conjugates, such as the deoxynivalenol (DON) glucosides formed during the fermentation of dough [4–6], binding to food-matrix components as shown for fumonisins [7], or the formation of degradation products, such as the norDON series [8].

Depending on the commodity, OTA undergoes different types of processing, such as baking, extrusion cooking, roasting, and fermentation. The reduction of OTA levels during these processes has been found to range between no impact and almost a 100% decrease [3,9–12]. A strong OTA decrease was observed during coffee roasting, where up to a 90% reduction was reported in most of the studies [13–17]. On the other hand, in some other coffee-roasting experiments, only a slight reduction of less than 12% was noted [18,19]. Several efforts have been made to identify the chemical reaction that leads to lower OTA levels in roasted coffee. Bittner et al. (2013) showed that OTA binds to polysaccharides of the coffee bean during roasting [20]. The high temperatures present during coffee roasting also lead to a decarboxylation to form decarboxy-ochratoxin A (DC-OTA) as well as to racemization of the phenylalanine moiety to yield 2′R-ochratoxin A (2′R-OTA, previously reported as 14(R)-OTA) and to the formation of ochratoxin α-amide (OTamide) as shown in Figure 1 [20–22]. A quantitative analysis of 15 coffee samples from the German market indicated that DC-OTA is only formed in minor amounts, while the concentration of 2′R-OTA was up to 0.63 µg/kg. A ratio of OTA and 2′R-OTA lower than 4:1 was determined [21]. OTamide was not analyzed in this study.

OTA has been shown to be nephrotoxic, hepatotoxic, teratogenic, and immunotoxic in various species and was classified by the International Agency for Research on Cancer (IARC) as possibly carcinogenic to humans (group 2B) in 1993 [23,24]. The mode of action is still not clear and controversially discussed [25,26]. In comparison to OTA, only very little information on the toxicity of the degradation product 2′R-OTA is available: 2′R-OTA shows in a Cell Counting Kit-8 (CCK-8) assay a 10-fold lower cytotoxic effect on IHKE-cells compared to OTA [21,27]. Several studies have shown that OTA is detectable in blood of humans and animals [28], with human plasma concentrations ranging roughly between 0.05 ng/mL and 5 ng/mL [29,30]. The thermal isomerization product 2′R-OTA is also of relevance, as it was recently found in humans in comparable levels as OTA [31].

Little is known on the kinetics of the isomerization of OTA to 2′R-OTA, and the minimum temperature required for the thermal conversion of OTA to 2′R-OTA has not been elucidated [21]. Depending on these data, bakery products and breakfast cereals might also be potential sources of 2′R-OTA exposure.

Herein, we report the analysis of 51 coffee samples from Germany, France, and Guatemala to extend the database on 2′R-OTA levels in food. Furthermore, the kinetics of the isomerization of OTA to 2′R-OTA were studied in a model system covering temperatures between 120 and 270 °C. As isomerization of OTA to 2′R-OTA has already been observed at 120 °C, further thermally treated food samples were analyzed on the occurrence of OTA and 2′R-OTA.

2. Results

2.1. Degradation of OTA in Model Heating Experiments

The isomerization of OTA to 2′R-OTA as well as the formation of other degradation products were studied in model heating experiments. To that end, OTA was heated at temperatures of 120, 150, 180, 210, 240, and 260 °C without solvent for 1 to 30 min, respectively. The observed degradation curves of OTA are shown in Figure 2. Subsequent analysis by HPLC-FLD for the known degradation products 2′R-OTA and DC-OTA resulted in the curves shown in Figure 3. At the lowest tested temperature of 120 °C, OTA remained almost stable over the entire heating period with only 3% 2′R-OTA formed within 30 min. After the same time, but heating at 150 °C, approximately 20% of OTA was converted to 2′R-OTA. At 180 °C, a fast racemization of OTA towards 2′R-OTA was observed with an equilibrium between both compounds reached after approximately 20 min. Above this temperature, racemization of OTA towards 2′R-OTA was achieved after 1 to 5 min of heating, followed by further degradation of both diastereomers. After 30 min at 210, 240, and 260 °C, only 80%, 35%, and 20%, respectively, of the sum of OTA and 2′R-OTA were detectable. Screening for DC-OTA and OTamide revealed only small quantities of less than 1% (data not shown), and were not further considered.

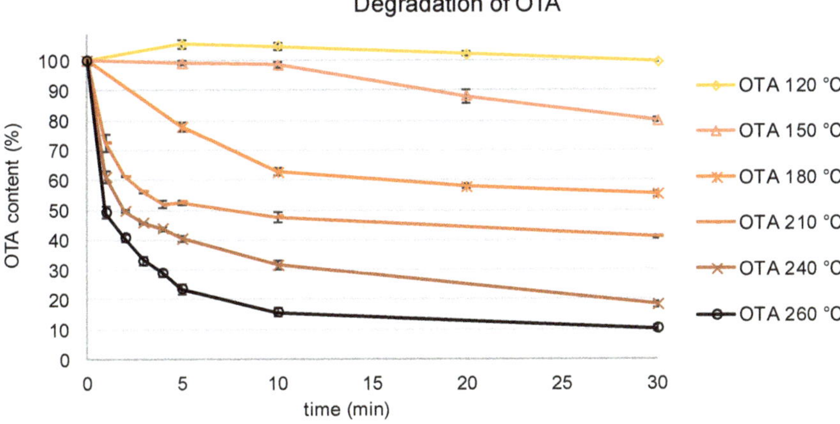

Figure 2. Degradation curves of ochratoxin A (OTA) at temperatures between 120 and 260 °C for 1–30 min.

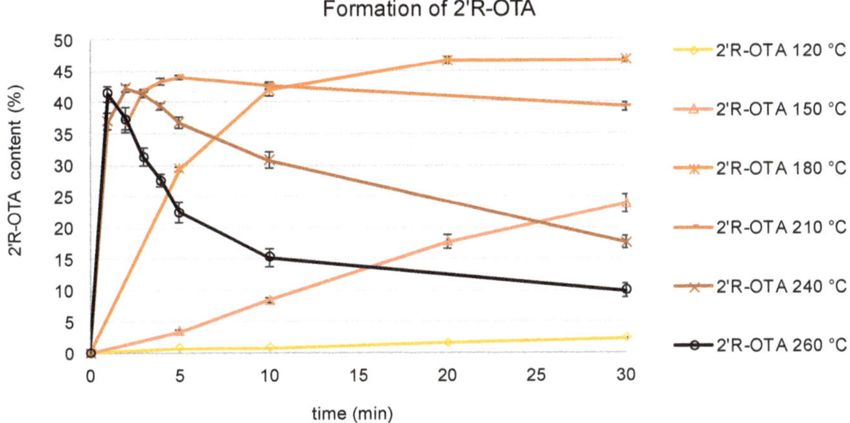

Figure 3. Formation of 2′R-ochratoxin A (2′R-OTA) during heating of OTA at temperatures between 120 and 260 °C for 1–30 min.

2.2. Coffee Powder: Current Situation of OTA and 2′R-OTA Content

The published dataset on 2′R-OTA contamination in food and the OTA:2′R-OTA ratio is limited to data from the analysis of 15 coffee samples in 2008 [21]. Therefore, in order to extend the database and to evaluate the current situation, we analyzed in total a set of 51 coffee samples from France, Germany, and Guatemala. Among the 14 commercially available roasted coffee samples from the French and German market, five were categorized as espresso powder and nine samples as coffee powder packages. Three of the coffee samples were labelled as organically grown and two as decaffeinated. The results are presented in Figure 4. In all coffee samples except two from Guatemala, OTA was quantifiable in a range between 0.26 and 28.4 µg/kg with a mean contamination of 1.98 µg/kg. No sample from Europe, but three samples from Guatemala (Figure 4, coffee samples 1-3) showed OTA levels above the legal limit of 5 µg/kg OTA set by the European Union [32]. Among these three, two samples exceeded the limit by factors of 5 and 4, respectively.

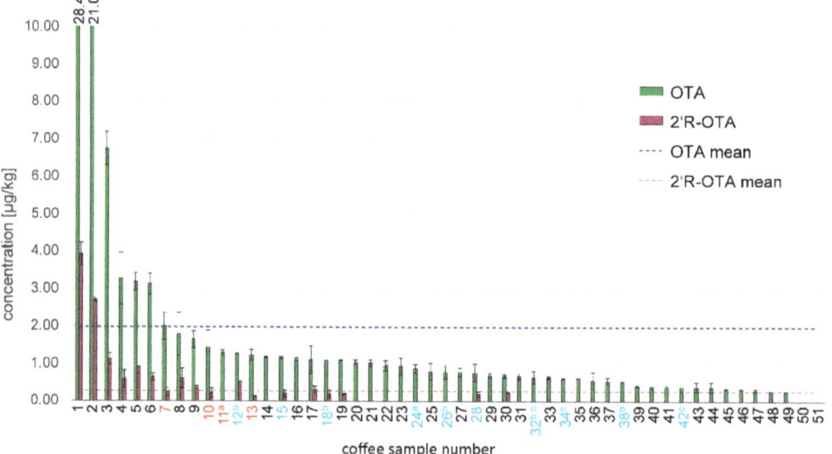

Figure 4. OTA and 2′R-OTA contents of the coffee powder samples from Germany (red), Guatemala (black), and France (blue). [a] decaffeinated, [b] espresso, [c] organically grown. The standard deviation is given for duplicate analysis.

In two decaffeinated coffee samples (Figure 4, samples 11 and 24), OTA concentrations of 1.30 and 0.88 µg/kg were detected. Interestingly, the three analyzed organic coffee powders, including one espresso sample (Figure 4, samples 32, 34, and 42), with OTA concentrations of 0.64, 0.62, and 0.38 µg/kg were among the coffee samples with the lowest detected OTA levels. The four other analyzed espresso samples (Figure 4, samples 12, 18, 26, and 38) contained 1.27, 1.09, 0.79, and 0.51 µg/kg OTA, which was comparable to the OTA contents of the other coffee samples from Europe.

The isomer 2′R-OTA was quantifiable in 18 of the 51 coffee samples (35%) with a mean concentration of 0.27 µg/kg. The highest 2′R-OTA level of 3.9 µg/kg was determined in the coffee sample from Guatemala containing 28 µg/kg OTA (Figure 4, sample 1). Among the European coffee samples, an espresso coffee powder (Figure 3, sample 12) with a 2′R-OTA concentration of 0.52 µg/kg and an OTA level of 1.27 µg/kg showed the highest 2′R-OTA contamination and, furthermore, the highest OTA:2′R-OTA ratio of 2.5:1. However, there was no indication that espresso coffee production favors 2′R-OTA formation, as the other analyzed espresso coffee samples contained no, or a low amount of, 2′R-OTA and showed OTA:2′R-OTA ratios between 4:1 and 10:1. In both decaffeinated coffee samples (Figure 3, samples 11 and 24), the 2′R-OTA concentration was below the limit of quantification (LOQ). Taking all coffee samples into account that were positive for 2′R-OTA, a OTA:2′R-OTA ratio between 10:1 and 2.5:1 with an average OTA:2′R-OTA-ratio of 5.5:1 was observed.

2.3. Evaluation of Other Thermally Processed Food Materials as Sources of 2′R-OTA Exposure

The data from the heating experiment with pure OTA suggest that, at a comparatively low temperature of 120 °C, a slow isomerization of OTA to 2′R-OTA can be observed. Such temperatures can be reached during different kinds of baking processes, and the conditions might be sufficient for the formation of significant quantities of 2′R-OTA. To that end, different thermal processed food samples were screened (Table 1). The choice of the analyzed food samples was based on thermal processing conditions and the overall contribution to OTA exposure [1]. Consequently, 30 samples of cocoa (8) and cereal products (22) were analyzed for OTA and 2′R-OTA. Details on the samples are given in Table 1.

Table 1. Results of a survey screening for the presence of OTA and 2′R-OTA in thermally processed food. (n: numbers of food samples; n.d.: not detectable).

Group	Food Sample		n	OTA (LOD, LOQ)	2′R-OTA (LOD, LOQ)
Cocoa Products		beans (roasted and unroasted)	2	1 sample > LOD	n.d.
		nibs (roasted and unroasted)	2	n.d.	n.d.
		Powder	2	1 sample > LOD	n.d.
		chocolate cream	2	1 sample > LOD	n.d.
Cereals	coffee like	instant malt coffee powder	5	4 samples > LOD 1 sample: 0.62 ± 0.04 µg/kg	3 samples > LOD 1 sample: 0.22 ± 0.02 µg/kg
		malt coffee powder	1	1 sample > LOD	1 sample > LOD
	expanded	puffed wheat	1	n.d.	n.d.
		rye puffed waffles	1	n.d.	n.d.
		rice puffed waffles	2	n.d.	n.d.
	Roasted	popcorn	2	n.d.	n.d.
		Breakfast cereals	2	1 samples > LOD	n.d.
		coloring malt	5	1 samples > LOD	n.d.
	baking	pumpernickel	4	3 samples > LOD 1 sample: 0.11 ± 0.02 µg/kg	1 sample > LOD

Generally, OTA contamination of the analyzed food samples was low, resulting in only two samples with OTA levels above the LOQ and 13 samples above the limit of detection (LOD). However, despite the low OTA contamination, six food samples were found to be positive for 2′R-OTA. In particular, coffee surrogates, such as instant malt coffee powder and malt coffee powder, were all found to be positive for this compound. With an OTA concentration of 0.62 ± 0.04 µg/kg and a 2′R-OTA concentration of 0.22 ± 0.02 µg/kg, comparable racemization rates as for roasted coffee were observed. Additionally, one bread sample (pumpernickel) contained detectable amounts of 2′R-OTA. Pumpernickel is a long-term heated bread. To produce this type of bread, a loaf of rye is baked for 16–24 h at an oven temperature of approximately 110 °C.

3. Discussion

The thermal instability of the mycotoxin OTA has been reported for several food processing technologies. However only for roasted coffee have the degradation products 2′R-OTA and DC-OTA been described and quantified so far, and in a limited number of samples. In this study, it was shown in model heating experiments that a slow racemization of OTA to 2′R-OTA occurs at temperatures as low as 120 °C. Above these temperatures, a fast racemization of OTA was observed, with 25% 2′R-OTA being formed after 30 min at 150 °C, after 5 min at 180 °C, and after 1 min or less at 210 °C and above. Longer heating periods at temperatures of 240 °C and above resulted in a fast degradation of OTA and 2′R-OTA, which can be explained by other reactions, such as pyrolysis or polymerization. Compared with previously reported data, these results confirm the importance of 2′R-OTA as the main OTA degradation product. A slow conversion of OTA to 2′R-OTA occurring at temperatures as low as 120 °C has not yet been reported [21,22].

Thus, food samples, processed at temperatures far lower than those applied for coffee roasting, might be additional sources of 2′R-OTA exposure. However, to date, 2′R-OTA has only been detected in blood samples from 34 coffee drinkers in concentrations between 0.021 and 0.411 ng/mL (mean: 0.21 ± 0.066 ng/mL) compared to 0.071–0.383 ng/mL (mean: 0.11 ± 0.093 ng/mL) for OTA. An average OTA:2′R-OTA ratio of 2:1 was determined and in some cases, the concentration of 2′R-OTA even exceeded that of OTA. No 2′R-OTA was detected in the set of 14 samples from non-coffee drinkers, suggesting that 2′R-OTA is predominantly present in roasted coffee. Nevertheless, the number of non-coffee drinkers participating in that study was low, and no correlation between the 2′R-OTA levels and overall coffee consumption was observed, making further sources of 2′R-OTA plausible [31,33].

Screening of a set of 51 coffee samples confirmed the role of roasted coffee as the key source of 2′R-OTA exposure, with 2′R-OTA levels of up to 3.9 µg/kg in highly contaminated coffee from Guatemala and up to 0.52 µg/kg in European coffee samples. No systematic differences between espresso coffee and other coffee with respect to the 2′R-OTA levels could be observed. For decaffeinated coffee, moderate OTA levels were detected, but no 2′R-OTA in concentrations above the LOQ. Although no information on the OTA content before roasting was available, the reported reductive effect of decaffeination on OTA levels seems to be limited [34]. These results are in good agreement with previously reported 2′R-OTA levels but also show that, in certain samples, such as one specific espresso powder, relatively high 2′R-OTA levels and OTA:2′R-OTA ratios as low as 2.5:1 can occur. In a previous study, for roasted coffee, OTA:2′R-OTA ratios between 10:1 and 4:1 (mean: 5:1) were determined from a set of 15 coffee samples. The variation of these ratios between different coffee samples also suggested a dependency of 2′R-OTA formation on the roasting process. Traditionally, green beans are roasted for between 8 and 20 min at temperatures between 160 and 240 °C using a drum roaster, a hot-air roaster, or a combination of both systems [35]. Oliveira et al. (2013) reported a correlation between roasting level and OTA degradation, while Castellanos-Onorio et al. (2011) observed for the two roasting techniques (drum roaster and hot-air roaster) a similar OTA reduction. In both cases, the formation or 2′R-OTA was unfortunately not studied [16,17]. In contrast to the traditional temperature regimes, to increase throughput, some companies have established a coffee-roasting process based on high temperatures of around 400 °C and roasting times of less than one minute [36].

To confirm the results from the model experiments, other thermally processed food samples, known to contribute to OTA exposure, were analyzed for the presence of 2′R-OTA, the most abundant thermal degradation product of OTA. Besides coffee, quantities of 2′R-OTA were detected in malt coffee as well as in traditionally baked rye bread (Pumpernickel). The latter is of special importance as it confirms the results from the model experiments, showing that OTA can be converted to 2′R-OTA at low temperatures between 100 and 120 °C. However, it has to be considered that this type of bread has a minimum baking time of 16 h, which makes a comparison with other bread and bakery products rather difficult. Other analyzed food samples contained no detectable amounts of 2′R-OTA but were also low for OTA.

The most relevant sources of OTA exposure are cereals, followed by coffee, cocoa, meat, and wine. For the cereal products biscuits and muesli, baking times are usually short and temperatures inside the product are mostly at 100 °C or below [37]. For roasted and expanded cereals, the situation is different; however, due to the low availability of industrially processed and naturally contaminated products, we were not able to prove whether a racemization of OTA during these processes may occur. Considering other typical food-processing procedures, wine can be excluded as potential source of 2′R-OTA due to low fermentation and processing temperatures. In the case of meat, temperatures above 100 °C are rarely reached in the inner part of the product (e.g., sausage). A different situation might be the manufacturing of canned meat, such as corned beef, where temperatures of up to 121 °C are applied for sterilization. Here, additional studies investigating the racemization of OTA in aqueous solutions are necessary. Thus, there is a need for more data on the occurrence of 2′R-OTA in these foods and on the toxic properties of this compound to allow for an adequate risk assessment.

4. Materials and Methods

Methanol (MeOH), acetonitrile (ACN), and toluene were obtained in gradient grade from Fisher Scientific (Schwerte, Germany). NaCl, formic acid, hexane, and Na_2HCO_3 were in pro analysi (p.a.)

quality from Merck KGaA (Darmstadt, Germany). Potato dextrose agar (PDA), potato dextrose broth, and KH_2PO_4 were from Carl Roth (Karlsruhe, Germany), Xylene in p.a. quality was obtained from Honeywell (Seelze, Germany), $NaHCO_3$ in p.a. quality was obtained from Grüssing (Filsum, Germany), and KCl in p.a. quality was obtained from VWR (Langenfeld, Germany). Purified water of ASTM type 1 quality was prepared with a Purelab Flex 2 system from Veolia Water Technologies (Celle, Germany).

4.1. Biosynthesis of Standards

OTA was isolated from cultures of *Aspergillus westerdijkiae* BFE 1115, provided by the Max Rubner Institute (Karlsruhe, Germany), which were activated for 2 days in liquid potato dextrose broth at room temperature and then incubated for two weeks at 27 °C on autoclaved durum wheat adjusted to a water content of 62.5% (*w/w*) and supplemented with 2.5% sodium chloride (*w/w*). OTA was extracted from the durum wheat cultures using tBME containing 0.5% formic acid, purified by liquid–liquid extraction with water at different pH levels and silica column chromatography with the solvent system toluene/tBME/formic acid (8/1.5/0.5 (*v/v/v*)). Finally, residual impurities were removed by crystallization from xylene/hexane (7/3 (*v/v*)). The purity of OTA was >99% as determined by HPLC-UV (220 nm) and NMR. 2'R-OTA, DC-OTA, OTα-amide, d_5-OTA, and d_5-2'R-OTA were prepared in-house as described elsewhere [21,22].

4.2. Model Heating Experiments with OTA

For the model heating experiments, a stock solution with a concentration of 274 µg/mL in acetonitrile was prepared. Aliquots of the solution (36.5 µL, 10 µg) were transferred into 1.5 mL vials and the solvent was evaporated to dryness under a stream of nitrogen at 40 °C. The dried thin film of OTA was heated for 1–30 min at temperatures of 120, 150, 180, 210, 240, and 260 °C, respectively. Subsequently, the samples were dissolved in 1 mL methanol/water/formic acid (63/37/0.15, (*v/v/v*)) and analyzed by HPLC-FLD for OTA and 2'R-OTA.

4.3. Sample Collection

European coffee samples were obtained from retail markets in France and Germany (and produced by industrial coffee roasting companies). The coffee samples originating from Guatemala were collected from small local markets and produced by local coffee roasters. Cocoa beans and nibs were provided as a gift from August Storck KG, Berlin, Germany. Other food samples reported in Table 1 were commercial products bought from retail markets in Germany.

4.4. Sample Preparation

Food samples were analyzed as described in literature with slight modifications [21,38]. To 5.00 g of homogenized ground coffee sample, 100 mL methanol/3% $NaHCO_3$-solution (1/1, *v/v*) and 50 µL of a solution containing 100 ng/mL d_5-OTA and 50 ng/mL d_5-2'R-OTA in methanol were added. The mixture was extracted for three minutes using an Ultra-Turrax T25 mixer (IKA, Staufen, Germany) at a rotation speed of 9500 min^{-1}. The obtained suspension was filtered through a 150 mm 3 HW folded filter (Sartorius-Stedim Biotech, Göttingen, Germany) and 5.00 mL of the filtrate were diluted with 45 mL phosphate-buffered saline (PBS) pH 7.4 (8 g NaCl, 1.2 g Na_2HPO_4, 0.2 g KCl, and 0.2 g KH_2PO_4 dissolved in 1 L H_2O) before purification using an OchraTest WB (VICAM, available via Klaus Ruttmann, Hamburg, Germany) immunoaffinity column (IAC). After loading the IAC with the sample, the column was washed with 10 mL PBS and 10 mL water. OTA, 2'R-OTA, and DC-OTA were eluted with 2 mL methanol according to the protocol of the IAC manufacturer. The eluate was evaporated to dryness under a stream of nitrogen at 40 °C and reconstituted in 250 µL methanol/water/formic acid (60/40/0.1, *v/v/v*).

4.5. Recovery Rate

For the determination of the recovery rate, a coffee sample containing only traces of OTA and 2'R-OTA was fortified with three different concentrations of OTA (0.5, 2, and 8 µg/kg) and 2'R-OTA

(0.4, 0.7, and 1.4 µg/kg) before analysis. To that end, aliquots of a standard solution of OTA and 2′R-OTA in acetonitrile were added to the homogenized sample and the solvent was allowed to evaporate for 2 h before extraction. The recovery rates were determined in duplicate and were 104.4 ± 4.0%, 102.9 ± 4.5%, and 103.9 ± 3.8% for 0.5, 2, and 8 µg/kg OTA and 106.0 ± 0.0%, 109.9 ± 11.2%, and 99.4 ± 3.8% for 0.4, 0.7, and 1.4 µg/kg 2′R-OTA, respectively.

4.6. Calibration

For HPLC-MS/MS experiments, an external seven-point calibration of OTA and 2′R-OTA with internal standards d_5-OTA (1 ng/mL) and d_5-2′R-OTA (0.5 ng/mL) in a concentration range from 0.1 to 10 ng/mL was used for quantification. The limit of detection (LOD) and the limit of quantification (LOQ) were determined by the signal-to-noise ratios 3 and 9, respectively. For OTA and 2′R-OTA, the LOQ was 0.1 ng/mL and the LOD was 0.03. For the HPLC-FLD experiments, a five-point calibration of OTA, 2′R-OTA (concentration range from 0.10 to 10.0 ng/mL), DC-OTA, and OTα-amide (concentration range from 0.05 to 2.00 ng/mL) was used for quantification. Samples exceeding the calibration range were diluted with methanol/water/formic acid (60/40/0.1, *v/v/v*) by an appropriate factor and reanalyzed. For all samples, calibration curves with correlation coefficient $r^2 > 0.95$ were calculated.

4.7. HPLC-MS/MS

An Agilent 1100 series HPLC (Agilent Technologies, Waldbronn, Germany) was coupled with an API 4000 QTRAP mass spectrometer (Sciex, Darmstadt, Germany) operated in electrospray ionization (ESI) positive mode with an ionization voltage of 5500 V. Data acquisition and quantification was carried out with the Analyst 1.6.2 software (Sciex). Chromatographic separation was achieved on a Nucleodur C18 Isis (150 × 2.0 mm; 5 µm) column with a 5 x 2 mm guard column of the same material (Macherey-Nagel, Düren, Germany) using a binary gradient at a flow rate of 0.3 mL/min. Methanol containing 0.1% formic acid was used as solvent A and water containing 0.1% formic acid as solvent B. The following linear gradient was used: 0 min 60% A, 1 min 60% A, 10 min 100% A, 10 min 100% A. The injection volume was 50 µL. The mass spectrometer was operated in selected reaction monitoring mode (SRM) with Gas 1 (nebulizer) set to 35 psi and Gas 2 (drying gas) set 350 °C and 45 psi. Nitrogen served as the Curtain gas (20 psi). The following transition reactions were monitored for 100 ms each (declustering potential (DP), collision energy voltage (CE), and collision cell exit potential (CXP) are given in brackets): OTA and 2′R-OTA: quantifier m/z 404.2→239.1 (DP 60 V, CE 33 V, CXP 14 V) and qualifier m/z 404.2→221.0 (DP 60 V, CE 50 V, CXP 15 V); d_5-OTA and d_5-2′R-OTA: quantifier m/z 409.2→239.1 (DP 60 V, CE 33 V, CXP 14 V), and quantifier m/z 409.2→221.0 (DP 60 V, CE 50 V, CXP 15 V).

4.8. HPLC-FLD

Samples of the degradation experiments were analyzed with HPLC Germany coupled to a fluorescence detector (X-LC, FP-2020 Plus, Jasco GmbH, Groß-Umstadt). Separation was achieved at 40 °C on a ReproSil-Pur C18-AQ (150 × 4.0 mm; 3 µm) column (Dr. Maisch GmbH, Ammerbuch, Germany) under isocratic conditions using a solvent mixture of methanol/water/formic acid (63/37/0.1, v/v/v) at a flow rate of 0.7 mL/min. The fluorescence detector was operated at an excitation wavelength of 330 nm and emission wavelength of 460 nm. The injection volume was 5 µL.

Author Contributions: Conceptualization, B.C., F.S. and H.-U.H.; investigation, F.S., O.T., V.H. and J.S.; resources, O.T. and H.-U.H.; writing—original draft preparation, F.S.; writing—review and editing, B.C. and H.-U.H.; supervision, B.C. and H.-U.H.; project administration, H.-U.H.; funding acquisition, B.C. and H.-U.H.

Funding: This research was funded by Deutsche Forschungsgemeinschaft (HU 730/10-2).

Acknowledgments: The authors thank Rolf Geisen for providing *A. westerdijkiae* BFE 1115.

Conflicts of Interest: The authors declare no conflict of interest.

References

1. European Commission. Assessment of dietary intake of ochratoxin A by the population of EU Member States. Report of the Scientific Cooperation, Task 3.2.7 Directorate-General Health and Consumer Protection, European Commission. Available online: https://ec.europa.eu/food/sites/food/files/safety/docs/cs_contaminants_catalogue_ochratoxin_task_3-2-7_en.pdf (accessed on 5 June 2019).

2. Karlovsky, P.; Suman, M.; Berthiller, F.; De Meester, J.; Eisenbrand, G.; Perrin, I.; Oswald, I.P.; Speijers, G.; Chiodini, A.; Recker, T.; et al. Impact of food processing and detoxification treatments on mycotoxin contamination. *Mycotoxin Res.* **2016**, *32*, 179–205. [CrossRef] [PubMed]

3. Caridi, A.; Sidari, R.; Pulvirenti, A.; Meca, G.; Ritieni, A. Ochratoxin A adsorption phenotype: An inheritable yeast trait. *J. Gen. Appl. Microbiol.* **2012**, *58*, 225–233. [CrossRef] [PubMed]

4. Vidal, A.; Sanchis, V.; Ramos, A.J.; Marín, S. Stability of DON and DON-3-glucoside during baking as affected by the presence of food additives. *Food Add. Contam. A* **2018**, *35*, 529–537. [CrossRef]

5. Vidal, A.; Marín, S.; Morales, H.; Ramos, A.J.; Sanchis, V. The fate of deoxynivalenol and ochratoxin A during the breadmaking process, effects of sourdough use and bran content. *Food Chem. Toxicol.* **2014**, *68*, 53–60. [CrossRef] [PubMed]

6. Wu, Q.; Kuča, K.; Humpf, H.U.; Klímová, B.; Cramer, B. Fate of deoxynivalenol and deoxynivalenol-3-glucoside during cereal-based thermal food processing: a review study. *Mycotoxin Res.* **2017**, *33*, 79–91. [CrossRef] [PubMed]

7. Seefelder, W.; Knecht, A.; Humpf, H.U. Bound fumonisin B-1: Analysis of fumonisin-B-1 glyco and amino acid conjugates by liquid chromatography-electrospray ionization-tandem mass spectrometry. *J. Agric. Food Chem.* **2003**, *51*, 5567–5573. [CrossRef] [PubMed]

8. Bretz, M.; Beyer, M.; Cramer, B.; Knecht, A.; Humpf, H.-U. Thermal degradation of the Fusarium mycotoxin deoxynivalenol. *J. Agric. Food Chem.* **2006**, *54*, 6445–6451. [CrossRef] [PubMed]

9. Scudamore, K.A.; Banks, J.; Macdonald, S.J. Fate of ochratoxin A in the processing of whole wheat grains during milling and bread production. *Food Addit. Contam.* **2003**, *20*, 1153–1163. [CrossRef] [PubMed]

10. Scudamore, K.A.; Banks, J.N.; Guy, R.C.E. Fate of ochratoxin A in the processing of whole wheat grain during extrusion. *Food Addit. Contam.* **2004**, *21*, 488–497. [CrossRef]

11. Castells, M.; Pardo, E.; Ramos, A.J.; Sanchis, V.; Marín, S. Reduction of ochratoxin A in extruded barley meal. *J. Food Prot.* **2006**, *69*, 1139–1143. [CrossRef] [PubMed]

12. Lee, H.J.; Kim, S.; Suh, H.J.; Ryu, D. Effects of explosive puffing process on the reduction of ochratoxin A in rice and oats. *Food Control* **2019**, *95*, 334–338. [CrossRef]

13. Blanc, M.; Pittet, A.; Munoz-Box, R.; Viani, R. Behavior of ochratoxin A during green coffee roasting and soluble coffee manufacture. *J. Agric. Food Chem.* **1998**, *46*, 673–675. [CrossRef] [PubMed]

14. Suarez-Quiroz, M.; de Louise, B.; Gonzalez-Rios, O.; Barel, M.; Guyot, B.; Schorr-Galindo, S.; Guiraud, J.-P. The impact of roasting on the ochratoxin A content of coffee. *Int. J. Food. Sci. Technol.* **2005**, *40*, 605–611. [CrossRef]

15. Van der Stegen, G.H.D.; Essens, P.J.M.; van der Lijn, J. Effect of roasting conditions on reduction of ochratoxin A in coffee. *J. Agric. Food Chem.* **2001**, *49*, 4713–4715. [CrossRef] [PubMed]

16. Castellanos-Onorio, O.; Gonzalez-Rios, O.; Guyot, B.; Fontana, T.A.; Guiraud, J.P.; Schorr-Galindo, S.; Durand, N.; Suárez-Quiroz, M. Effect of two different roasting techniques on the ochratoxin A (OTA) reduction in coffee beans (*Coffea arabica*). *Food Control* **2011**, *22*, 1184–1188. [CrossRef]

17. Oliveira, G.; da Silva, D.M.; Alvarenga Pereira, R.G.; Paiva, L.C.; Prado, G.; Batista, L.R. Effect of different roasting levels and particle sizes on ochratoxin A concentration in coffee beans. *Food Control* **2013**, *34*, 651–656. [CrossRef]

18. Studer-Rohr, I.; Dietrich, D.R.; Schlatter, J.; Schlatter, C. The occurrence of ochratoxin A in coffee. *Food Chem. Toxicol.* **1995**, *33*, 341–355. [CrossRef]

19. Tsubouchi, H.; Yamamoto, K.; Hisada, K.; Sakabe, Y.; Udagawa, S. Effect of roasting on ochratoxin A level in green coffee beans inoculated with *Aspergillus ochraceus*. *Mycopathologia* **1987**, *97*, 111–115. [CrossRef]

20. Bittner, A.; Cramer, B.; Humpf, H.U. Matrix binding of ochratoxin A during roasting. *J. Agric. Food Chem.* **2013**, *61*, 12737–12743. [CrossRef]

21. Cramer, B.; Königs, M.; Humpf, H.-U. Identification and in vitro cytotoxicity of ochratoxin A degradation products formed during coffee roasting. *J. Agric. Food Chem.* **2008**, *56*, 5673–5681. [CrossRef]

22. Bittner, A.; Cramer, B.; Harrer, H.; Humpf, H.U. Structure elucidation and in vitro cytotoxicity of ochratoxin α amide, a new degradation product of ochratoxin A. *Mycotoxin Res.* **2015**, *31*, 83–90. [CrossRef] [PubMed]

23. IARC. *Some Naturally Occurring Substances: Food Items and Constituents, Heterocyclic Aromatic Amines and Mycotoxins*; Monographs on the Evaluation of Carcinogenic Risks to Humans; IARC: Lyon, France, 1993.

24. Malir, F.; Ostry, V.; Pfohl-Leszkowicz, A.; Malir, J.; Toman, J. Ochratoxin A: 50 years of research. *Toxins* **2016**, *8*, 191. [CrossRef] [PubMed]

25. Ostry, V.; Malir, F.; Toman, J.; Grosse, Y. Mycotoxins as human carcinogens—the IARC Monographs classification. *Mycotoxin Res.* **2017**, *33*, 65–73. [CrossRef] [PubMed]

26. Rottkord, U.; Röhl, C.; Ferse, I.; Schulz, M.C.; Rückschloss, U.; Gekle, M.; Schwerdt, G.; Humpf, H.U. Structure–activity relationship of ochratoxin A and synthesized derivatives: Importance of amino acid and halogen moiety for cytotoxicity. *Arch. Toxicol.* **2017**, *91*, 1461–1471. [CrossRef] [PubMed]

27. Cramer, B.; Harrer, H.; Nakamura, K.; Uemura, D.; Humpf, H.-U. Total synthesis and cytotoxicity evaluation of all ochratoxin A stereoisomers. *Bioorg. Med. Chem.* **2010**, *18*, 343–347. [CrossRef] [PubMed]

28. Valenta, H. Chromatographic methods for the determination of ochratoxin A in animal and human tissues and fluids. *J. Chromatogr. A* **1998**, *815*, 75–92. [CrossRef]

29. Duarte, S.C.; Pena, A.; Lino, C.M. Human ochratoxin A biomarkers-from exposure to effect. *Crit. Rev. Toxicol.* **2011**, *41*, 187–212. [CrossRef]

30. Studer-Rohr, I.; Schlatter, J.; Dietrich, D.R. Kinetic parameters and intraindividual fluctuations of ochratoxin A plasma levels in humans. *Arch. Toxicol.* **2000**, *74*, 499–510. [CrossRef]

31. Cramer, B.; Osteresch, B.; Munoz, K.A.; Hillmann, H.; Sibrowski, W.; Humpf, H.U. Biomonitoring using dried blood spots: Detection of ochratoxin A and its degradation product 2′R-ochratoxin A in blood from coffee drinkers. *Mol. Nutr. Food Res.* **2015**, *59*, 1837–1843. [CrossRef]

32. European Commission. Commission Regulation (EC) No 1881/2006 of 19 December 2006 setting maximum levels for certain contaminants in foodstuffs. *Off. J. Eur. Union* **2006**, *L 364*, 5–24.

33. Viegas, S.; Osteresch, B.; Almeida, A.; Cramer, B.; Humpf, H.U.; Viegas, C. Enniatin B and ochratoxin A in the blood serum of workers from the waste management setting. *Mycotoxin Res.* **2018**, *34*, 85–90. [CrossRef] [PubMed]

34. Eggers, R.; Pietsch, A. Technology I: Roasting. Coffee—Recent developments. In *Coffee Recent Developments*; Clarke, R.J., Vietzthum, O.G., Eds.; Blackwell Science: London, UK, 2001; pp. 90–116.

35. Durand, N.; Duris, D.; Fontana, A.; Castellanos-Onorio, O.; Suarez-Quiroz, M.; Schorr-Galindo, S. Effects of post-harvest and roasting treatments on Ochratoxine A levels in coffee: A review paper. *Cahiers Agric.* **2013**, *22*, 195–201. [CrossRef]

36. Prince, S.; Kussin, R.; Fruhling, R.; Harris, M. Ultrafast Roasted Coffee. U.S. Patent 4,988,590, 29 January 1991.

37. Kuchenbuch, H.S.; Becker, S.; Schulz, M.; Cramer, B.; Humpf, H.-U. Thermal stability of T-2 and HT-2 toxins during biscuit- and crunchy muesli-making and roasting. *Food Add. Contam. A* **2018**, *35*, 2158–2167. [CrossRef] [PubMed]

38. Pittet, A.; Tornare, D.; Huggett, A.; Viani, R. Liquid chromatographic determination of ochratoxin A in pure and adulterated soluble coffee using an immunoaffinity column cleanup procedure. *J. Agric. Food Chem.* **1996**, *44*, 3564–3569. [CrossRef]

Article

The Influence of Processing Parameters on the Mitigation of Deoxynivalenol during Industrial Baking

David Stadler [1], Francesca Lambertini [2], Lydia Woelflingseder [3], Heidi Schwartz-Zimmermann [1], Doris Marko [3], Michele Suman [2], Franz Berthiller [1,*] and Rudolf Krska [1,4]

1   Institute of Bioanalytics and Agro-Metabolomics, Department of Agrobiotechnology (IFA-Tulln), University of Natural Resources and Life Sciences, Vienna (BOKU), Konrad-Lorenz-Str. 20, 3430 Tulln, Austria; david.stadler@boku.ac.at (D.S.); heidi.schwartz@boku.ac.at (H.S.-Z.); rudolf.krska@boku.ac.at or r.krska@qub.ac.uk (R.K.)
2   Barilla G. R. F.lli SpA, Advanced Laboratory Research, via Mantova 166, 43122 Parma, Italy; francesca.lambertini@barilla.com (F.L.); michele.suman@barilla.com (M.S.)
3   Department of Food Chemistry and Toxicology, University of Vienna, Währingerstraße 38, 1090 Vienna, Austria; lydia.woelflingseder@univie.ac.at (L.W.); doris.marko@univie.ac.at (D.M.)
4   Institute for Global Food Security, School of Biological Sciences, Queens University Belfast, University Road, Belfast BT7 1NN, Northern Ireland, UK
*   Correspondence: franz.berthiller@boku.ac.at; Tel.: +43-1-47654-97371

Received: 13 May 2019; Accepted: 30 May 2019; Published: 4 June 2019

Abstract: Deoxynivalenol (DON), a frequent contaminant of flour, can be partially degraded by baking. It is not clear: (i) How the choice of processing parameter (i.e., ingredients, leavening, and baking conditions) affects DON degradation and thus (ii) how much DON can be degraded during the large-scale industrial production of bakery products. Crackers, biscuits, and bread were produced from naturally contaminated flour using different processing conditions. DON degradation during baking was quantified with the most accurate analytical methodology available for this *Fusarium* toxin, which is based on liquid chromatography tandem mass spectrometry. Depending on the processing conditions, 0–21%, 4–16%, and 2–5% DON were degraded during the production of crackers, biscuits, and bread, respectively. A higher $NaHCO_3$ concentration, baking time, and baking temperature caused higher DON degradation. NH_4HCO_3, yeast, vinegar, and sucrose concentration as well as leavening time did not enhance DON degradation. In vitro cell viability assays confirmed that the major degradation product isoDON is considerably less toxic than DON. This proves for the first time that large-scale industrial baking results in partial detoxification of DON, which can be enhanced by process management.

Keywords: mycotoxins; trichothecenes; thermal degradation; decontamination; mass spectrometry; food processing; detoxification; design of experiment; LC-MS/MS

Key Contribution: Deoxynivalenol (DON) can be partially detoxified during the industrial production of bakery products. The detoxification of DON can be enhanced by process management.

---

1. Introduction

Deoxynivalenol (DON) is along with aflatoxins, fumonisins, ochratoxin A, and zearalenone, one of the main mycotoxins of significance for human disease [1]. DON frequently contaminates cereals and cereal-based products [2]. The population of the European Union (EU) is frequently exposed to DON mainly due to the consumption of bread and other bakery wares [2]. Therefore, the European Commission (EC) has set maximum levels for DON in milling products and bread/bakery wares of

750 µg/kg and 500 µg/kg, respectively [3]. This legislation follows the rationale that it is feasible to reduce DON from flour during the production of bakery products by 33%. This assumption is supported by the recent report on the risk associated with DON in food by the European Food Safety Authority (EFSA) [2]. From 2007 to 2014, EFSA gathered occurrence data for the concentration of DON present in different food groups. For milling products, bread and rolls, and fine bakery wares, 4609, 2837, and 975 data points, respectively, were collected from the EU member states. The mean and 95th percentile concentrations of DON were 20–30% lower in bread and bakery wares (bread and rolls, fine bakery wares) than in flour (milling products).

During baking, DON was found to degrade mainly to isoDON, but also to norDON B and norDON C [4–6]. norDON B and norDON C were found to be considerably less cytotoxic than DON [5]. isoDON was found to have much lower inhibitory potency on the ribosome than DON, which is strongly indicative of its overall lower toxicity [6].

The reduction of the DON concentration during baking can be due to formulation (i.e., the dilution effect due to mixing contaminated flour with non-contaminated ingredients) or due to degradation of DON. The reduction due to formulation can be easily calculated from the percentage of flour of the finished bakery product, assuming no additional sources of DON. The extent of the DON degradation during baking, however, was reported to be very variable [2,7,8]. This high variation can be explained by errors in the calculation of the amount of DON that was degraded (e.g., by not considering the dilution with other ingredients) and by not performing baking trials with a suitable number of replicates. Recently, we identified all degradation products of DON, formed during baking of crackers, biscuits, and bread, using a stable isotope labelling approach combined with high resolution mass spectrometry analysis. After quantifying DON and all of its formed degradation products, we were able to provide the first comprehensive mass balance for those commodities [6].

In this current study, DON degradation during the production of crackers, biscuits, and bread from flour fortified with DON was determined to be 6%, 5%, and 2%, respectively, based on the increase of degradation products. As this study was aimed at the elucidation of the degradation products that are formed during baking, the experiments were carried out under standard baking conditions. As a step forward, the influence of the processing conditions should be evaluated. Furthermore, experiments with naturally contaminated flour need to be conducted in order to reproduce the actual industrial starting conditions in place during the large-scale production of bakery products.

The influence of different processing parameters on DON degradation during the production of crackers, biscuits, bread, and rusks was determined in previous studies [9–12]. Statistical evaluation of the DON concentration showed that baking temperature, baking time, and the concentration of the raising agent $NaHCO_3$ influence the amount of DON in the final products. However, the statistical models as well as the determination of the DON degradation may have suffered from the high analytical variability of the determination of the DON concentration.

We hypothesized that DON degradation during industrial baking is influenced by the choice of processing parameter (i.e., ingredients, leavening, and baking conditions). Hence, the objectives of this study were to: (i) Produce bakery products (crackers, biscuits, and bread) from naturally contaminated flour using different processing conditions, (ii) determine DON degradation by LC-MS/MS based quantification of the DON degradation products and iii) develop a statistical model which shows the influence of the processing parameters on the DON degradation. Due to the comprehensive baking trials in combination with the highly sensitive LC-MS/MS analysis and the consideration of all formed degradation products, this study presents the most accurate report on the effect of industrial baking on DON degradation to date. Moreover, a first characterization of the cytotoxicity of isoDON, compared to DON, is presented.

2. Results

To study the main processing factors affecting the mitigation of DON during industrial baking, crackers, biscuits, and bread were chosen as representative commodities of the main bakery categories produced with very different technologies. A Design of Experiments (DoE) approach was applied to set up the experimental trials. In agreement with results previously obtained by our research group [9,10,12], the main technological parameters monitored were: (i) Baking conditions (i.e., time and temperature), (ii) pH modifying agents (NaHCO$_3$, NH$_4$HCO$_3$, and other minor ingredients) and (iii) leavening conditions (time, temperature, presence, and type of leavening agents). Each factor was modified within a range according to technological feasibility. Some conditions stressed the process over the organoleptic acceptability of the final product. These processing conditions were chosen to promote significant changes of the DON concentration in order to understand the impact of the applied processing conditions. In the present work, all the food commodities were produced at a pilot plant level. As a starting point of the DoE approach, the same processing parameters as those on common industrial processing lines were chosen. Figure 1 exemplifies the influence of the baking conditions on the final appearance of biscuits, bread and crackers.

Figure 1. Display of the influence of baking conditions on the appearance of biscuits (top), bread (middle) and crackers (bottom).

2.1. Incluence of Processing Parameters on DON Degradation during the Production of Bakery Products

For all three baking commodities, the results of the experimental trials were evaluated and presented in two sub-chapters. First, the sum of the DON degradation products, which is the measure for DON degradation, was shown as a bar graph. The individual experiments were listed according to the processing parameters that were found to be important for DON degradation based on a visual assessment. In this way, Figures 2, 4, and 6 show the DON degradation that was achieved under the specified processing conditions ranked from high to low DON degradation. These bar graphs give a first impression of the DON degradation and the influence of the processing parameter for the individual commodities. Second, to confirm and refine the conclusions of the visual assessment, a DoE model was built using the MODDE software tool. The outcome of this statistical analysis is visualized by an effects plot which displays the change of the sum of the DON degradation products when a processing factor is varied from its lowest to its highest value (Figures 3, 5, and 7, top). The data set obtained from the DoE model was used to build a prediction model for the influence of the process parameter that led to a statistically significant increase of DON degradation. These predictive factor plots show how DON degradation can be enhanced by the design of the production process (Figures 3, 5, and 9, bottom). To allow for the easy comparability of the results between the three baking commodities, the same scaling was chosen for the individual graphs.

2.1.1. Biscuits

Biscuits, from the technological point of view, represent a quite simple bakery preparation characterized by a baking step and the use of raising agents. In order to monitor the impact of these two factors on the DON content in the final products, a very basic recipe was chosen, containing only flour, oil, sugar, and water, with the addition of $NaHCO_3$ and NH_4HCO_3 as leavening agents. Several batches of biscuits were prepared at different conditions according to the DoE inputs based on the Screening Interaction model. In the present study, the processing parameters $NaHCO_3$ concentration, NH_4HCO_3 concentration, baking time, baking temperature, and sugar content were varied (Table S1). Due to the different production conditions, DON degradation varied between 4–16% (Figure 2).

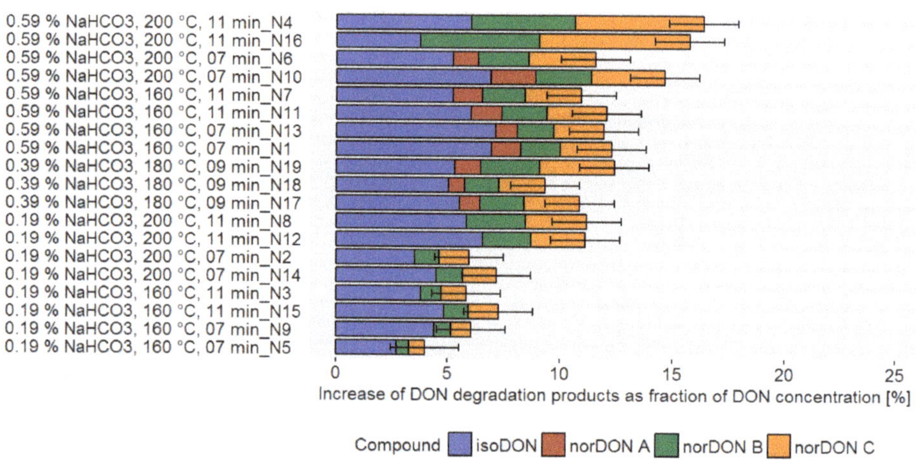

Figure 2. Increase of the deoxynivalenol (DON) degradation products isoDON and norDONs A-C during the production of biscuits using different processing conditions. The increase was determined as a ratio of the molal concentration of the DON degradation products to the molal concentration of DON resulting from the natural contamination of the flour. The experimental trials were listed according to the $NaHCO_3$ concentration, baking temperature, and baking time. Error bars represent the process standard deviation.

A higher DON degradation was observed in batches with a higher concentration of $NaHCO_3$, baking temperature, and baking time. The variation of the processing parameter also led to a changed ratio of the degradation products with the main degradation products being norDON B and C, instead of isoDON, for experiments with a higher baking temperature and baking time. norDON A, which was not found to be a degradation product in our previous study [6], was found only at processing conditions that lead to higher DON degradation.

The DoE model confirmed that the main factors affecting DON degradation were: (i) The quantity of $NaHCO_3$ used, (ii) the baking temperature, and (iii) the baking time (Figure 3, top). In addition, the statistical model indicates that a synergistic effect of baking temperature and time exists. Changes in NH_4HCO_3 and sucrose concentration did not significantly influence DON degradation products.

The prediction model for the influence of the processing parameter on DON degradation revealed that by increasing the $NaHCO_3$ concentration from 0.2 to 0.6%, DON degradation is increased by 5% (Figure 3, bottom). An increase in baking temperature and time is predicted to enhance DON degradation by about 2%.

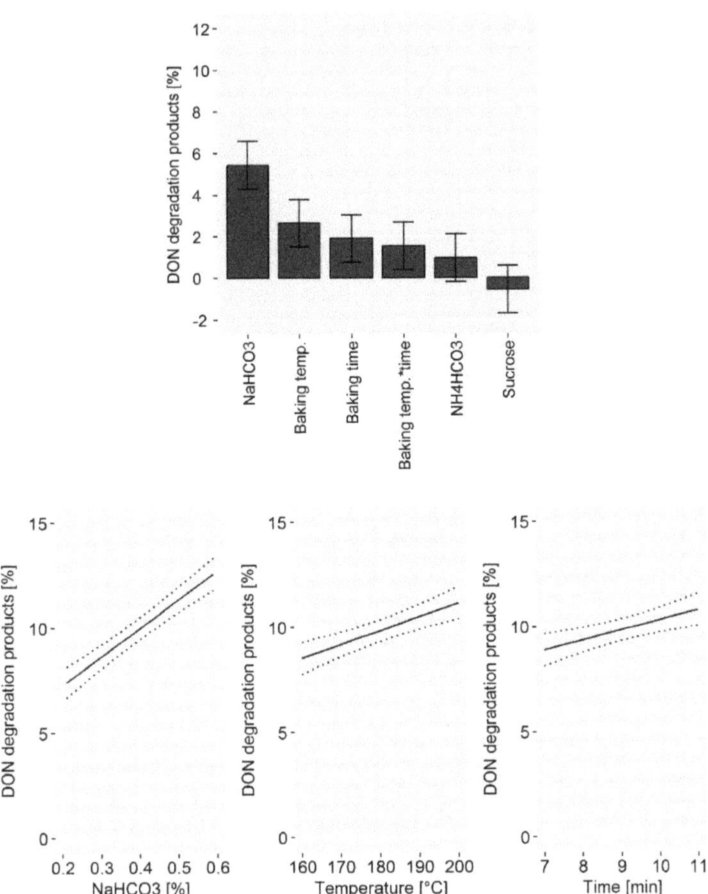

Figure 3. Top: Effects plot, which shows the change of the sum of the DON degradation products when a processing factor is varied from its lowest to its highest value and all other factors are kept at their averages, which was obtained for the design of the experiment (DoE) data set in pilot-scale biscuit making experiments. Error bars represent the confidence interval corresponding to a 95% confidence level. "*" between two processing factors indicates a synergistic effect, which cannot be explained solely by addition of degradation caused by the two parameters individually. Bottom: Predictive factor effect plots, which show the influence of the following processing parameters on the deoxynivalenol (DON) degradation during the biscuit production: (i) $NaHCO_3$ (left), (ii) baking temperature (middle), and (iii) baking time (right). The dotted lines represent the confidence interval corresponding to a 95% confidence level.

2.1.2. Bread

Bread is surely the most representative and most studied commodity of soft bakery products. Its production workflow implies dough fermentation before baking. Both fermentation and subsequent baking can be managed for time and temperature. The fermentation phase can also be modulated for relative humidity conditions, yeast, bakery improvers, raising agents, and pH. As a consequence, the number of factors to investigate for assessing potential DON mitigation effects was quite high. Therefore, in comparison to biscuits and crackers, more pilot plant trials had to be carried out. Based on literature data and our experience, the following variables were taken into account for the DoE approach: Baking temperature, baking time, leavening time, yeast, sucrose, and cider vinegar concentration. Each factor was varied within a range defined according to technological feasibility (Table S4). Depending on the processing parameter, DON degradation varied from 2–5% (Figure 4).

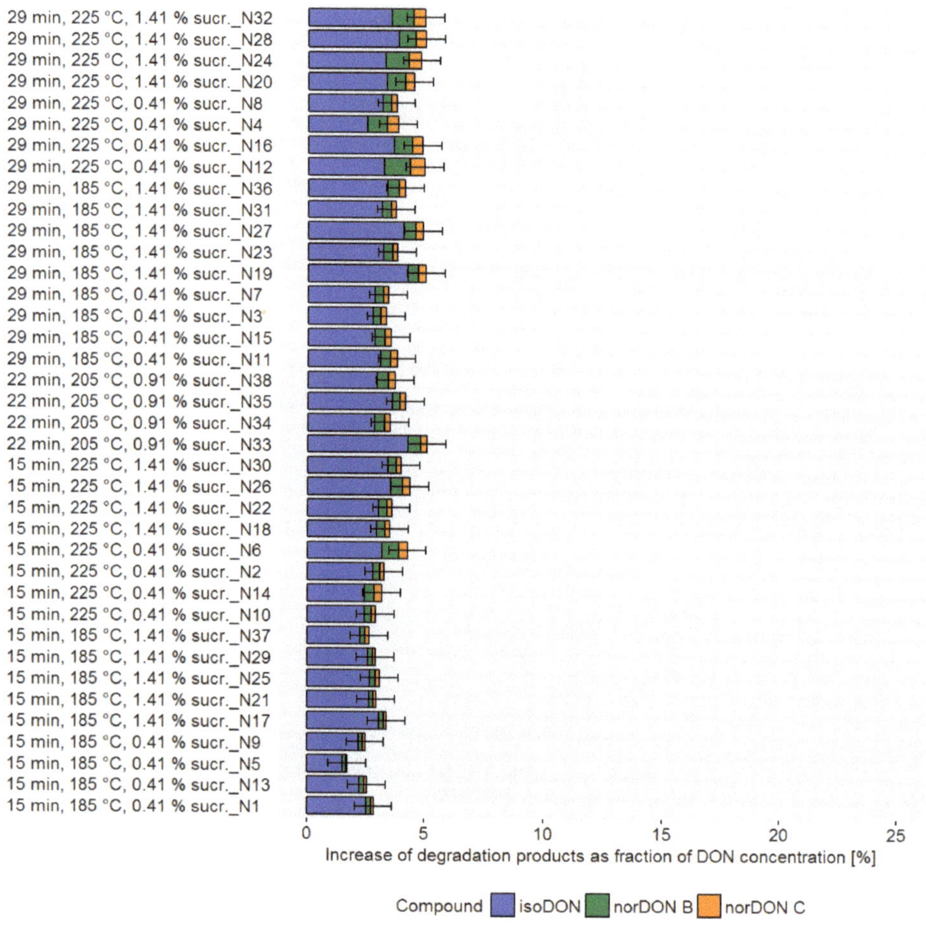

Figure 4. Increase of the deoxynivalenol (DON) degradation products isoDON and norDONs B-C during the production of bread using different processing conditions. The increase was determined as a ratio of the molal concentration of the DON degradation products to the molal concentration of DON resulting from the natural contamination of the flour. The experimental trials were listed according to baking temperature, baking time, and sucrose concentration. Error bars represent the process standard deviation.

During the bread production, the achievable DON degradation was found to be low compared to the biscuit and cracker production. Although the influence was small, we found that the baking time and temperature as well as the sucrose concentration impacted the DON concentrations. The DoE model confirmed that the main factors influencing DON degradation were: (i) Baking time, (ii) baking temperature, and (iii) sucrose concentration (Figure 5, top). Changes in cider vinegar and yeast concentration as well as the leavening time did not influence DON degradation. The prediction model confirmed that changes in the processing parameter have only a minor effect of approximately 2% on DON degradation (Figure 5, bottom).

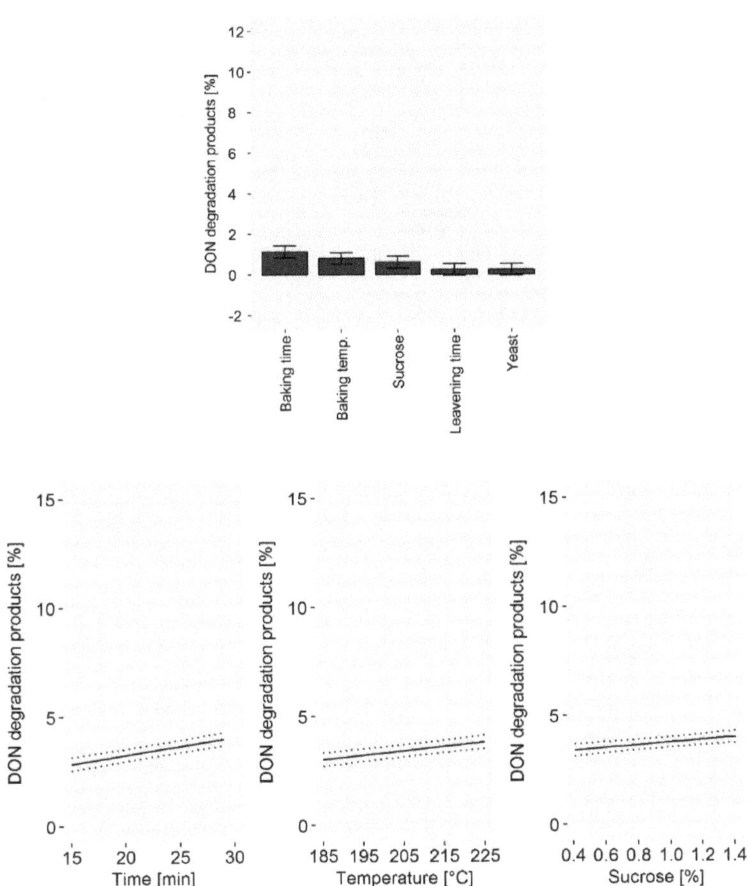

Figure 5. Top: Effects plot, which shows the change of the sum of the DON degradation products when a processing factor is varied from its lowest to its highest value and all other factors are kept at their averages, which was obtained for the design of the experiment (DoE) data set in pilot-scale bread making experiments. Error bars represent the confidence interval corresponding to a 95% confidence level. Bottom: Predictive factor effect plots, which show the influence of the following processing parameters on the deoxynivalenol (DON) degradation during the biscuit production: (i) Baking time (left), (ii) baking temperature (middle), and (iii) sucrose concentration (right). The dotted lines represent the confidence interval corresponding to a 95% confidence level.

2.1.3. Crackers

Crackers were chosen as being representative of a complex baking commodity involving different technological aspects and combining several factors which were demonstrated to affect DON mitigation in the previous experiments carried out on biscuits and bread. Among the cracker making parameters, the following conditions were considered as factors in the experimental design: Baking time, baking temperature, acidic mother content, and $NaHCO_3$ concentration. For this trial, each factor was varied within a range defined according to technological feasibility (Table S7). During the cracker production, the DON degradation varied from 0–21%, depending on the choice of processing parameters (Figure 6).

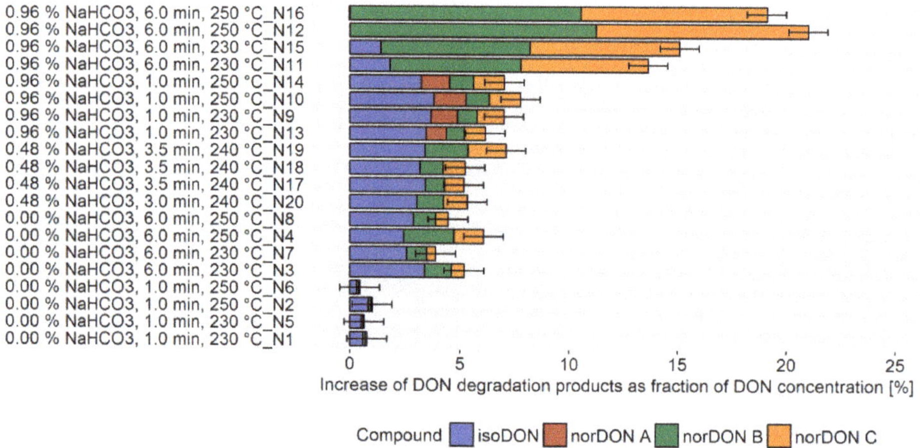

Figure 6. Increase of the deoxynivalenol (DON) degradation products isoDON and norDONs A-C during the production of crackers using different processing conditions. The increase was determined as a ratio of the molal concentration of the DON degradation products to the molal concentration of DON resulting from the natural contamination of the flour. The experimental trials were listed according to the $NaHCO_3$ concentration, baking time, and baking temperature. Error bars represent the process standard deviation.

A high DON degradation was observed in experiments with a high $NaHCO_3$ concentration, baking time, and baking temperature. Similar to the biscuit production, the variation of the processing parameter caused a change in the ratio of the degradation products. The DoE model revealed that $NaHCO_3$ concentration and baking time were the main factors affecting DON degradation (Figure 7, top). In addition, the statistical model indicated that a synergistic effect of baking time and $NaHCO_3$ concentration exists. The acidic mother concentration, leavening time, and baking temperature did not affect DON degradation. The prediction model revealed the potential DON reduction that can be achieved by process management (Figure 7, bottom). An increase of the $NaHCO_3$ concentration from 0 to 1% is predicted to increase DON degradation by 10%. Similarly, the increase of baking time from 1 to 6 min is predicted to increase DON degradation by 10%.

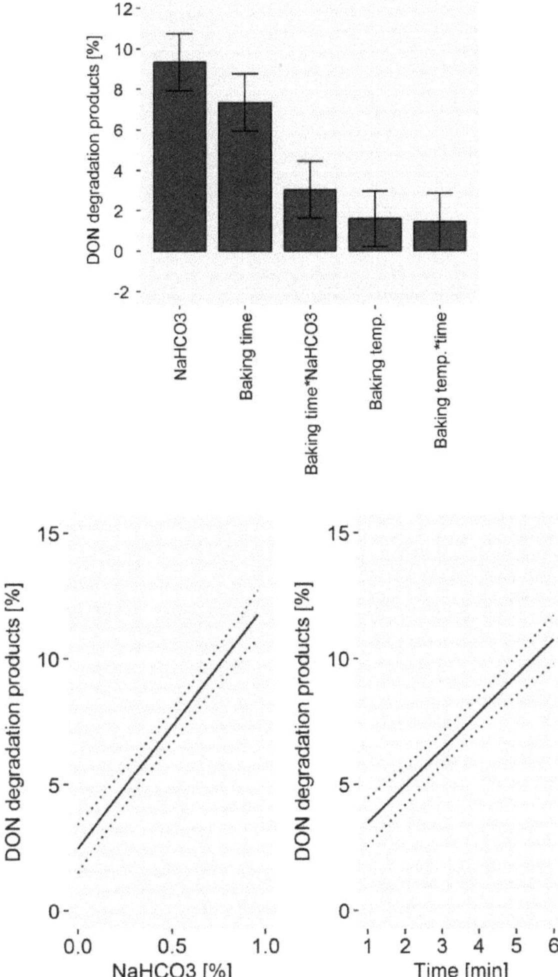

Figure 7. Top: Effects plot, which shows the change of the sum of the DON degradation products when a processing factor is varied from its lowest to its highest value and all other factors are kept at their averages, which was obtained for the design of the experiment (DoE) data set in pilot-scale bread making experiments. Error bars represent the confidence interval corresponding to a 95% confidence level. "*" between two processing factors indicates a synergistic effect, which cannot be explained solely by addition of degradation caused by the two parameters individually. Bottom: Predictive factor effect plots, which show the influence of the following processing parameters on the deoxynivalenol (DON) degradation during the biscuit production: (i) NaHCO3 concentration (left) and (ii) baking time (right). The dotted lines represent the confidence interval corresponding to a 95% confidence level.

2.2. Cytotoxic Effects of isoDON

The cytotoxic potential of isoDON in direct comparison to DON was evaluated in two different human colon cell lines (Figure 8).

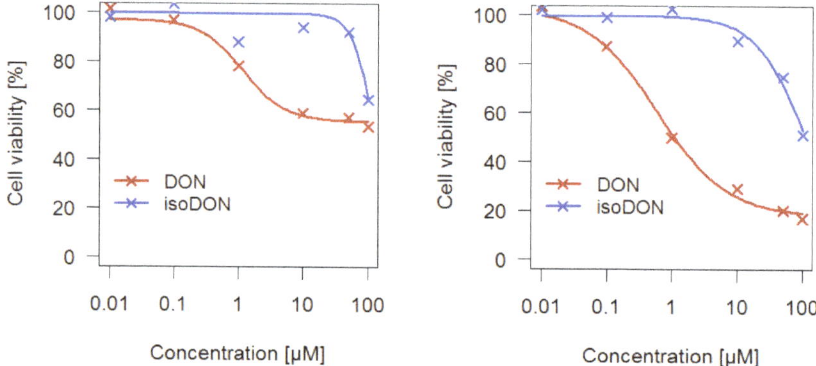

Figure 8. Cell viability of human colorectal adenocarcinoma cells (HT-29) (left) and non-tumorigenic human colon epithelial cells (HCEC) (right) after 24 h of incubation with different concentrations of deoxynivalenol (DON) and isoDON. The concentration at which cell viability was reduced by 30% (IC_{30}) was calculated from a dose response curve fitted to the individual data points ($\times$).

Usually, the half maximal inhibitory concentration (IC_{50}) is used to compare the potency of two substances. As the highest tested isoDON concentrations did not reduce cell viability by 50% under all tested conditions, IC_{30} values where cell viability was reduced by 30% were used instead. For the human colorectal adenocarcinoma cell line HT-29 the following IC_{30} values were determined: 1.9 µM (DON) and 91 µM (isoDON), whereas for the non-tumorigenic human colon epithelial cells (HCEC) the following IC_{30} values were identified: 0.4 µM (DON) and 53 µM (isoDON). In both cell viability assays, isoDON was much less potent (factor 48 and 133 for HT-29 and HCEC, respectively) in reducing cellular viability compared to DON.

3. Discussion

3.1. Influence of Different Processing Parameters on DON Degradation in Bakery Products

3.1.1. The pH Value of the Dough

As DON has been reported to be unstable in alkaline solutions, the pH of the dough is clearly an important processing factor to consider when designing a production process that is optimized for a high DON degradation [13]. For the production of biscuits and crackers, the pH of the dough was regulated primarily by chemical raising agents. We found that the type of the chemical raising agent as well as its concentration was crucial regarding DON degradation. Whereas the use of $NaHCO_3$ led to higher DON degradation, the use of NH_4HCO_3 did not influence DON concentration. This can be explained by the different pH values of the dough resulting from the different chemical nature of the raising agents. Gökmen et al. prepared cookies from two different doughs containing 0.7% $NaHCO_3$ and NH_4HCO_3, respectively [14]. When $NaHCO_3$ was used, the pH was initially 8.5 and increased slightly during baking to pH 9. In the absence of acidic compounds, $NaHCO_3$ is thermally converted to Na_2CO_3, H_2O and CO_2 (Equation (1)). As CO_3^{2-} is a stronger base than HCO_3^{-}, the thermal degradation of $NaHCO_3$ leads to an increase of the ph. The use of NH_4HCO_3 as a leavening agent led to a decrease of the pH from 8 in the dough to 6–7 in the baked cookies. The decrease of pH can be explained by the degradation of NH_4HCO_3 to NH_3, CO_2 and H_2O, which evaporate during baking (Equation (2)).

$$2\,NaHCO_3 \rightarrow Na_2CO_3 + H_2O + CO_2 \tag{1}$$

$$NH_4HCO_3 \rightarrow NH_3 + H_2O + CO_2 \tag{2}$$

Besides chemical raising agents, further ingredients were shown to affect the pH of the dough. High pH values due to the presence of $NaHCO_3$ were only observed when sucrose was used as sugar. At a higher pH, sucrose was not hydrolyzed and thereby did not cause a change in pH [14]. When glucose was used, it was partially hydrolyzed to fructose resulting in a decrease of the pH from 8.5 in the dough to 6–7 in the baked cookies. This can be rationalized by the report of Feng et al. who calculated the pKa values for glucose and fructose [15]. Fructose was shown to be considerably more acidic (pK_a 12) compared to glucose (pK_a 14). In the presence of an acid, $NaHCO_3$ was converted to CO_2 which caused a decrease of the pH (Equation (3)).

$$NaHCO_3 + H^+ \rightarrow Na^+ + H_2O + CO_2 \tag{3}$$

For the biscuit production, our results confirm the outcome of a previous study, where the $NaHCO_3$ concentration was the main factor that enhanced DON degradation [9]. For the cracker production, our results contradict a previous report in which baking temperature and time were found to be of higher importance compared to the $NaHCO_3$ concentration. However, in the previous study, fewer experimental trials were carried out and the accuracy of the analytical methodology was lower.

3.1.2. Baking Conditions

For all three commodities, the baking conditions (i.e., temperature and/or time) influenced DON degradation. However, for the production of biscuits and crackers, baking conditions were less important compared to the pH of the dough. For the production of crackers and bread, baking time was found to be more important than baking temperature. The contrary was observed for biscuit production. The reason for this observation might relate to differences in moisture content and surface to volume ratio of the individual baking commodities.

3.1.3. Ratio of the DON Degradation Products

For the production of biscuits and crackers, the ratio of the DON degradation products shifted from isoDON to norDON B and C in the experiments that led to high DON degradation. This supports the proposed mechanism, that isoDON is further converted to norDON B and C [13]. Some baking conditions led to the formation of norDON A. Most likely, the reason why norDON A was not detected in a previous study was a higher limit of detection (LOD) and limit of quantification (LOQ) for norDON A compared to isoDON, norDON B and C in the analytical method used [6]. Interestingly, the formation of norDON A was not observed in the experiments with the highest DON degradation (i.e., biscuits: N4, N16; crackers: N11, N12, N15, N16) in total. This might have been due to an initial formation of isoDON, which was further degraded into norDON A and finally to norDON B or norDON C.

3.2. Toxicity of the DON Degradation Products

For a comprehensive assessment of the health risks of eating bakery products produced from DON contaminated flour, the toxicity of all degradation products has to be determined. So far, norDONs A–C were shown to be considerably less cytotoxic than DON [5]. In comparison to DON, isoDON was found in in vitro translation assays to have a considerably lower potency (factor 94 and 60 in wheat and rabbit ribosomes, respectively) to inhibit translation, known to be the major mechanism of toxicity of trichothecenes [6]. As the 60S subunit binding site of ribosomes is the main molecular target of DON, the results of the translation assay were indicative of considerably lower cytotoxicity [16,17]. However, various aspects, such as different cellular absorption, changes in distribution, and especially differences in fitting into the pockets of the ribosomal A-site due to changes in polarity might lead to significant differences in cytotoxicity of isoDON and DON. Therefore, the cytotoxic effects of isoDON in comparison to DON were evaluated.

Cytotoxic Effects of isoDON

Performing cell viability assays in two different human colon cell lines, we found that isoDON induced significantly less cytotoxicity compared to DON in HCEC and HT-29 cells. In Figure 9, the structures of DON and isoDON, showing structural differences at the carbon atoms C-7 to C-10 of the trichothecene backbone, are displayed.

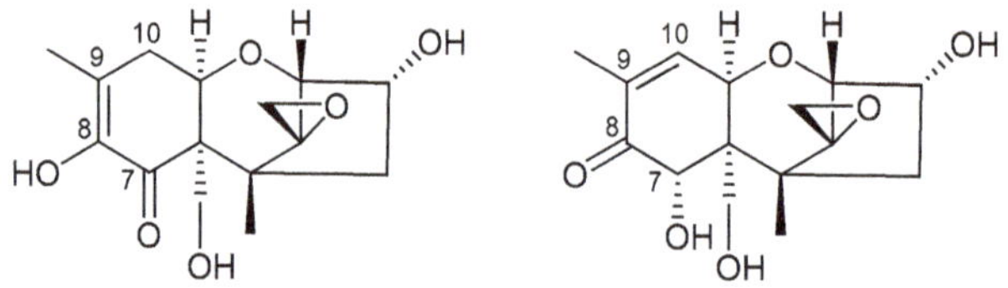

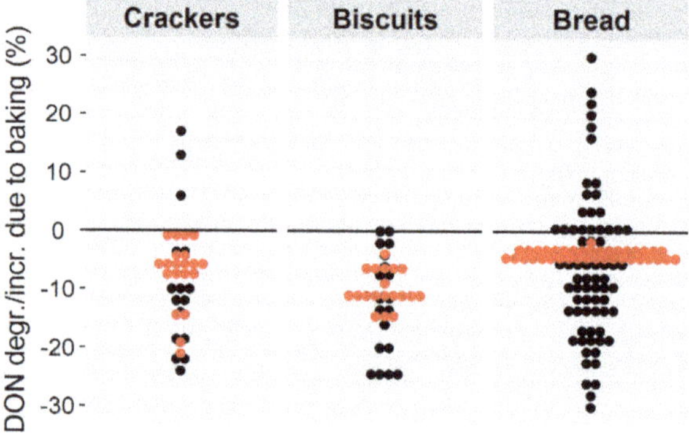

Figure 9. Structure of iso-deoxynivalenol (isoDON) (left) and DON (right). The C-7 to C-10 atoms of the trichothecene backbone are highlighted.

Our findings are supported by structure activity relationships which have been conducted for various trichothecenes [18–20]. The absence of a double bond at the C-9 and C-10 position, which is the case for isoDON, led to a significant decrease in both inhibition of protein biosynthesis and cytotoxicity. Although the toxicity of DON is often ascribed to the epoxide moiety, we could provide additional proof that a double bond between C-9 and C-10 is an essential structural feature for ribosomal inhibition and thus, for trichothecene-induced toxicity.

4. Conclusions

Depending on the processing conditions, 0–21%, 4–16%, and 2–5% of DON were degraded during the production of crackers, biscuits, and bread, respectively. For biscuits and crackers, DON degradation can be increased by 10–20% by process management. For bread, process management can lead to a minor increase in DON degradation of about 3%. DON degradation was enhanced by a high $NaHCO_3$ concentration, baking time, and temperature. The processing parameters NH_4HCO_3, yeast, cider vinegar, acidic mother content, and sucrose concentration as well as leavening time did not influence DON concentration.

To put our findings in context to the results reported in recent literature, a graphical summary is presented in Figure 10.

Figure 10. Comparison between the deoxynivalenol (DON) degradation reported in recent literature studies [10–12,21–29] and the DON degradation values determined in this study under different processing conditions. Each dot represents one baking experiment.

Compared to previous studies found in the literature, we achieved a much narrower range of DON degradation, which was most pronounced for bread. The reason for that is most likely that we applied the currently most accurate analytical methodology to determine DON degradation [6].

To gain further knowledge about the toxicity of DON degradation products, the cytotoxicity of isoDON was evaluated by performing in vitro cell viability assays. isoDON was found to be considerably less cytotoxic (factor 48 and 133 for HT-29 and HCEC cells, respectively) compared to DON. As all DON degradation products are considerably less cytotoxic than DON, we conclude that DON degradation during baking results in a lower toxicity. Thus, for the first time, we have presented proof that baking under real industrial conditions can lead to a partial detoxification of DON.

5. Materials and Methods

5.1. Chemicals and Reagents

Acetonitrile (ACN, gradient grade) was purchased from VWR International GmbH (Vienna, Austria). Acetic acid (LC–MS gradient grade) was obtained from Sigma Aldrich (Vienna, Austria). In all experiments, ultra-pure water (purified by a Purelab Ultra system ELGA LabWater, Celle, Germany) was used. Liquid calibrant solutions of DOM-1, DON-3-Glc, DON, and U-[^{13}C$_{15}$]-DON were supplied by Romer Labs GmbH (Tulln, Austria). Reference standards of the DON degradation products isoDON and norDONs A, B and C were synthesized according to published procedures [5,6,21,30].

5.2. Preparation of Bakery Products

The food commodities used in the present study were produced from naturally contaminated flour according to the procedure previously described by our group by applying a suitable scaling up factor of the starting ingredients that fits with the production under pilot plant facilities [6].

Biscuits: The final dough (3000 g, 14% moisture content (MC)) was used to form the individual biscuits (approximately 11 g each) which consisted of 56 wt % flour, 40 wt % fat and sugar, and 4 wt % water.

Bread: The final dough (8000 g, 14% MC) was used to form the individual loafs of 450 g each (40% MC), that reached a final weight of 400 g and consisted of 65 wt % flour, yeast, oil, and salt and 35 wt % water after baking.

Crackers were shaped (approximately 10 g each) from the final dough (4127 g, 26% MC) and they consisted of 97 wt % flour, yeast, and malt extract and 3 wt % water.

5.3. Design of Experiment (DoE) Setup and Statistical Evaluation

Design of Experiments (DoE) was used to determine the influence of processing parameters on the degradation of DON and the increase of its degradation products during the baking of crackers, biscuits, and bread from naturally contaminated flour. The screening interaction model was chosen and processing parameters that we hypothesized to influence DON degradation were used as model quantitative factors. Table 1 provides a list of the processing parameters that were investigated in our study. For more detailed information on the processing parameters used in the individual experiments, the reader is referred to Tables S1, S4, and S7.

To focus the attention on the increase of DON degradation products during baking, multivariate data analysis was performed on the concentration (µg/kg) obtained by LC/MS-MS analysis in the finished products of the sum of DON degradation products. DoE set up and data elaboration were carried out by using MODDE software 11 (Umetrics, Umea, Sweden).

Table 1. Processing parameters that were investigated for their impact on the degradation of deoxynivalenol during the production of bakery products from naturally contaminated flour under pilot plant conditions.

Condition	Biscuits	Bread	Crackers
Sucrose concentration [%]	12.5 to 17.5	0.41 to 1.41	-
NH$_4$HCO$_3$ concentration [%]	0.21 to 0.61	-	-
NaHCO$_3$ concentration [%]	0.19 to 0.59	-	0.00 to 0.96
Vinegar concentration [%]	-	0.00 to 0.36	-
Yeast concentration [%]	-	0.83 to 1.33	-
Leavening time [%]	-	70 to 100	-
Acidity Mother [%]	-	-	0.65 to 1.65
Baking time [min]	7 to 11	15 to 29	1 to 6
Baking temperature [°C]	160 to 200	185 to 225	230 to 250
Total experiments	19	38	20

5.4. Analysis of the Samples

To determine DON degradation during the industrial baking process, the concentrations of DON and its degradation products (i.e., isoDON, norDON B, and norDON C) were analyzed in naturally contaminated flour and the products made thereof. Additionally, the concentration of the suspected DON degradation and conversion products DON-3-glucoside, DOM-1 and norDON A was monitored. Sample preparation and analysis are discussed in detail elsewhere [6]. In brief, 5.00 ± 0.01 g of ground and homogenized sample was extracted with 20.0 mL of ACN:H$_2$O, 84:16, *v:v*. A total of 200 µL of sample extract were dried down and reconstituted in 100 µL of water. Subsequently, the concentrated sample extracts were centrifuged at 10 °C and 4500 rpm for 10 min. Four µL aliquots of the concentrated sample extract were injected together with 0.4 µL of internal standard into the HPLC system. Each sample was worked-up once and injected in duplicate. The analytical method used for the analysis was based on LC-MS/MS and was validated for recovery, repeatability, LOD, and LOQ. All results were corrected for the recovery factor determined in the validation study [6]. The concentration of the analytes that was found in naturally contaminated flour and in the finished bakery products can be found in Table S2 and S3 (biscuits), Table S5 and S6 (bread), and Table S8 and S9 (crackers) of the Supplementary Material, respectively.

5.5. Calculation of the DON Degradation

The sum of the DON degradation products was shown to be a much more accurate measure of the DON degradation than the change of the DON content itself [6]. Therefore, the DON degradation was calculated based on the increase of its degradation products isoDON and norDONs A–C. As the flour used for the production of the bakery products was naturally contaminated with the degradation products at different concentration levels, the increase of the degradation products was calculated separately before being summed up to obtain the final DON degradation.

5.6. Calculation of the Increase of the DON Degradation Products Due to Baking

For the production of bakery products from naturally contaminated flour, the increase of the degradation products was calculated according to:

$$dilution\ factor = \frac{m_{flour}}{m_{bakery\ product}}$$

$$b_{analyte,dilution\ of\ nat.cont.\ flour} = b_{analyte,\ flour} * dilution\ factor$$

$$Change\ due\ to\ baking = \frac{b_{analyte,\ bakery\ product} - b_{analyte,dilution\ of\ nat.cont.\ flour}}{b_{DON,dilution\ of\ nat.cont.\ flour}}$$

First, the dilution factor, which describes the mass ratio of naturally contaminated flour present in the finished bakery product, was determined. The dilution factor was 0.97 (crackers), 0.56 (biscuits), and 0.65 (bread), respectively. Second, the theoretical molal concentration b (mol/kg) of the analyte resulting from the dilution of the flour by the addition of other ingredients ($b_{analyte,dilution\ of\ nat.cont.\ flour}$) was determined. Finally, the molal concentration of an analyte resulting from the natural contamination of the flour was subtracted from the molal concentration of the analyte in the bakery product ($b_{analyte,\ bakery\ product}$) and expressed as a fraction of the molal concentration of DON resulting from the natural contamination of the flour. The result was the fraction of DON that is converted to DON degradation products during the production of bakery products.

5.7. Estimation of the Process Standard Deviation

In each DoE, three center points were set, which means that three experiments were carried out under identical processing conditions: Experiment number 17, 18 and 19 (crackers), 17, 18 and 19 (biscuits) and 33, 34, 35 (bread), respectively. The process standard deviation was calculated for the sum of the change of the degradation products of the three replicates which were produced using the same processing parameter. The process standard deviation was 0.9%, 1.4%, and 0.8% for crackers, biscuits, and bread, respectively, and thus considerably higher than the analytical standard deviation for the determination of the sum of the DON degradation products which was in the range of 0.2–0.4% [6].

5.8. Human Cell Culture

Human colorectal adenocarcinoma cells, HT-29, were purchased from the German Collection of Microorganisms and Cell Cultures (DSMZ, Braunschweig, Germany). For cell culture Dulbecco's Modified Eagle's Medium (DMEM) supplemented with 10% (*v/v*) heat inactivated fetal calf serum and 50 U/mL penicillin and 50 µg/mL streptomycin was used. The non-tumorigenic human colon epithelial cell line, HCEC [31], was kindly provided by Prof. Jerry W. Shay (UT Southwestern Medical Center, Dallas, TX, USA). HCEC cells were cultivated in DMEM (high glucose) supplemented with 2% (*v/v*) Medium 199 (10×), 2% (*v/v*) cosmic calf serum, 20 mM 4-(2-hydroxyethyl)-1-piperazineethanesulfonic acid (HEPES), 50 µg/mL gentamicin, insulin-transferrin-selenium-G (10 µg/mL; 5.5 µg/mL; 6.7 µg/mL), recombinant human epidermal growth factor (18.66 ng/mL), and hydrocortisone (1 µg/mL). Both cell lines were cultivated in humidified incubators at 37 °C and 5% CO_2. Media, supplements, and further material for cell culture were purchased from GIBCO Invitrogen (Karlsruhe, Germany), Lonza Group Ltd. (Basel, Switzerland), Sigma-Aldrich Chemie GmbH (Munich, Germany), Sarstedt AG & Co. (Nuembrecht, Germany), and Fisher Scientific GmbH (Vienna, Austria).

5.9. Cytotoxicity of isoDON

Per well, 1500 HCEC or 5500 HT-29 cells were seeded in a 200 µL culture medium in 96-well plates and allowed to grow for 48 h. Then, the culture medium was withdrawn and cells were treated with different concentrations of isoDON and DON in the respective cell culture medium for 24 h. As a solvent control, 1% water (LC–MS grade) was incubated and 1% (*v/v*) Triton X-100 served as a positive control. After the treatment period, the medium was withdrawn and cells were rinsed once with phosphate buffered saline (100 µL/well) and incubated with DMEM containing 10% (*v/v*) alamarBlue® reagent (Invitrogen™ Life Technologies, Karlsruhe, Germany) for 75 min. Subsequently, fluorescence intensity was measured (excitation: 530 nm; emission: 600 nm) on a Cytation 3 Imaging Multi Mode Reader (BioTek, Bad Friedrichshall, Germany). For quantification, blank-values were subtracted and measured data were referred to solvent control. Data are presented as a mean of at least five independent experiments, each performed in technical triplicates.

Supplementary Materials: The following are available online at http://www.mdpi.com/2072-6651/11/6/317/s1, Table S1: Processing parameters used for the production of biscuits in different experiments (Exp. No.). Ingredients are given as the weight percentage of the dough; Table S2: Concentration of the analytes in the naturally contaminated flour, dilution factor by which the flour gets diluted in the final biscuits, and the resulting concentration (assuming 0% degradation). <LOQ: concentration was below the limit of quantification (LOQ) of the analytical methodology. Table S3: Concentration of deoxynivalenol (DON) and related compounds in the biscuit samples. <LOQ: concentration was below the limit of quantification (LOQ) of the analytical methodology; Table S4: Processing parameters used for the production of bread in different experiments (Exp. No.). Ingredients are given as the weight percentage of the dough; Table S5: Concentration of the analytes in the naturally contaminated flour, dilution factor by which the flour gets diluted in the final bread, and the resulting concentration (assuming 0% degradation). n.d. (not detected): Concentration was below the limit of quantification of the analytical methodology; Table S6: Concentration of deoxynivalenol (DON) and related compounds in bread samples. n.d. (not detected): Concentration was below the limit of quantification of the analytical methodology; Table S7: Processing parameters used for the production of crackers in different experiments (Exp. No.). Ingredients are given as the weight percentage of the dough; Table S8: Concentration of the analytes in the naturally contaminated flour, dilution factor by which the flour gets diluted in the final crackers, and the resulting concentration (assuming 0% degradation). n.d. (not detected): Concentration was below the limit of quantification of the analytical methodology; Table S9: Concentration of deoxynivalenol (DON) and related compounds in cracker samples. n.d. (not detected): Concentration was below the limit of quantification of the analytical methodology.

Author Contributions: Conceptualization, D.S., F.L., M.S., F.B. and R.K.; methodology, D.M., M.S. and R.K.; data curation, D.S., F.L. and L.W.; toxicity study, L.W and D.M.; visualization, D.S. and F.L.; writing-original draft preparation, D.S., F.L. and L.W.; writing—review and editing, all authors; supervision, D.M., M.S., F.B. and R.K.; project administration and funding acquisition, D.M., M.S. and R.K.

Funding: This project has received funding from the European Union's Horizon 2020 research and innovation programme under grant agreement No. 678012 for the MyToolBox project, also covering the APC. We also acknowledge financial support from the Austrian Science Fund (FWF) via the special research project Fusarium (F3718).

Acknowledgments: The authors would like to thank the colleagues for the pilot plant trials execution support: Dante Catellani, Marta Filomeno, Gennaro Cangiano, Sergio Cerasti and Umberto Tanzi.

Conflicts of Interest: The authors declare no conflict of interest. The funders had no role in the design of the study; in the collection, analyses, or interpretation of data; in the writing of the manuscript, or in the decision to publish the results.

References

1. Pitt, J.I.; Wild, C.P.; Baan, R.A.; Gelderblom, W.C.A.; Miller, J.D.; Riley, R.T.; Wu, F. *Improving Public Health through Mycotoxin Control*; IARC Scientific Publications Series, No. 158; International Agency for Research on Cancer: Lyon, France, 2012.

2. Knutsen, H.K.; Alexander, J.; Barregård, L.; Bignami, M.; Brüschweiler, B.; Ceccatelli, S.; Cottrill, B.; Dinovi, M.; Grasl-Kraupp, B.; Hogstrand, C.; et al. Risks to human and animal health related to the presence of deoxynivalenol and its acetylated and modified forms in food and feed. *EFSA J.* **2017**, *15*. [CrossRef]

3. European Commission. European Commission Commission Regulation (EC) No 1881/2006 of 19 December 2006 setting maximum levels for certain contaminants in foodstuff. *Off. J. Eur. Union* **2006**, *364*, 5–24.

4. Greenhalgh, R.; Gilbert, J.; King, R.R.; Blackwell, B.A.; Startin, J.R.; Shepherd, M.J. Synthesis, characterization, and occurrence in bread and cereal products of an isomer of 4-deoxynivalenol (vomitoxin). *J. Agric. Food Chem.* **1984**, *32*, 1416–1420. [CrossRef]

5. Bretz, M.; Beyer, M.; Cramer, B.; Knecht, A.; Humpf, H.U. Thermal degradation of the Fusarium mycotoxin deoxynivalenol. *J. Agric. Food Chem.* **2006**, *54*, 6445–6451. [CrossRef] [PubMed]

6. Stadler, D.; Lambertini, F.; Bueschl, C.; Wiesenberger, G.; Hametner, C.; Schwartz-zimmermann, H.; Hellinger, R.; Sulyok, M.; Lemmens, M.; Schuhmacher, R.; et al. Untargeted LC-MS based 13C labelling provides a full mass balance of deoxynivalenol and its degradation products formed during baking of crackers, biscuits and bread. *Food Chem.* **2019**, *279*, 303–311. [CrossRef] [PubMed]

7. Schaarschmidt, S.; Fauhl-Hassek, C. The fate of mycotoxins during the processing of wheat for human consumption. *Compr. Rev. Food Sci. Food Saf.* **2018**, *17*, 556–593. [CrossRef]

8. Khaneghah, M.A.; Fakhri, Y.; Sant'Ana, A.S. Impact of unit operations during processing of cereal-based products on the levels of deoxynivalenol, total a fl atoxin, ochratoxin A, and zearalenone: A systematic review and meta-analysis. *Food Chem.* **2018**, *268*, 611–624. [CrossRef] [PubMed]

9. Generotti, S.; Cirlini, M.; Šarkanj, B.; Sulyok, M.; Berthiller, F.; Dall'Asta, C.; Suman, M. Formulation and processing factors affecting trichothecene mycotoxins within industrial biscuit-making. *Food Chem.* **2017**, *229*, 597–603. [CrossRef] [PubMed]

10. Suman, M.; Manzitti, A.; Catellani, D. A design of experiments approach to studying deoxynivalenol and deoxynivalenol-3-glucoside evolution throughout industrial production of wholegrain crackers exploiting LC-MS/MS techniques. *World Mycotoxin J.* **2012**, *5*, 241–249. [CrossRef]

11. Generotti, S.; Cirlini, M.; Malachova, A.; Sulyok, M.; Berthiller, F.; Dall'Asta, C.; Suman, M. Deoxynivalenol & deoxynivalenol-3-glucoside mitigation through bakery production strategies: Effective experimental design within industrial rusk-making technology. *Toxins* **2015**, *7*, 2773–2790. [PubMed]

12. Bergamini, E.; Catellani, D.; Dall'Asta, C.; Galaverna, G.; Dossena, A.; Marchelli, R.; Suman, M. Fate of fusarium mycotoxins in the cereal product supply chain: The deoxynivalenol (DON) case within industrial bread-making technology. *Food Addit. Contam.* **2010**, *27*, 677–687. [CrossRef] [PubMed]

13. Young, J.C.; Blackwell, B.A.; ApSimon, J.W. Alkaline degradation of the mycotoxin 4-deoxynivalenol. *Tetrahedron Lett.* **1986**, *27*, 1019–1022. [CrossRef]

14. Gökmen, V.; Açar, Ö.Ç.; Serpen, A.; Morales, F.J. Effect of leavening agents and sugars on the formation of hydroxymethylfurfural in cookies during baking. *Eur. Food Res. Technol.* **2008**, *226*, 1031–1037. [CrossRef]

15. Feng, S.; Bagia, C.; Mpourmpakis, G. Determination of proton affinities and acidity constants of sugars. *J. Phys. Chem. A* **2013**, *117*, 5211–5219. [CrossRef] [PubMed]

16. Cundliffe, E.; Cannon, M.; Davies, J. Mechanism of inhibition of eukaryotic protein synthesis by trichothecene fungal toxins. *Proc. Natl. Acad. Sci. USA* **1974**, *71*, 30–34. [CrossRef] [PubMed]

17. Ueno, Y.; Nakajima, M.; Sakai, K.; Ishii, K.; Sato, N. Comparative toxicology of trichothec mycotoxins: Inhibition of protein synthesis in animal cells. *J. Biochem.* **1973**, *74*, 285–296.

18. Anderson, D.W.; Black, R.M.; Lee, C.G.; Pottage, C.; Rickard, R.L.; Sandford, M.S.; Webber, T.D.; Williams, N.E. Structure-Activity Studies of Trichothecenes: Cytotoxicity of Analogues and Reaction Products Derived from T-2 Toxin and Neosolaniol. *J. Med. Chem.* **1989**, *32*, 555–562. [CrossRef]

19. Wu, Q.; Dohnal, V.; Kuca, K.; Yuan, Z. Trichothecenes: Structure-toxic activity relationships. *Curr. Drug Metab.* **2013**, *14*, 641–660. [CrossRef] [PubMed]

20. Wei, C.; Mclaughlin, S. Structure-function relationship in the 12,13-epoxytrichothecenes. *Biochem. Biophys. Commun.* **1974**, *57*, 838–844. [CrossRef]

21. Scudamore, K.A.; Hazel, C.M.; Patel, S.; Scriven, F. Deoxynivalenol and other Fusarium mycotoxins in bread, cake, and biscuits produced from uk-grown wheat under commercial and pilot scale conditions. *Food Addit. Contam.* **2009**, *26*, 1191–1198. [CrossRef]

22. Voss, K.A.; Snook, M.E. Stability of the mycotoxin deoxynivalenol (DON) during the production of flour-based foods and wheat flake cereal. *Food Addit. Contam.* **2010**, *27*, 1694–1700. [CrossRef] [PubMed]

23. Vidal, A.; Sanchis, V.; Ramos, A.J.; Marín, S. Thermal stability and kinetics of degradation of deoxynivalenol, deoxynivalenol conjugates and ochratoxin A during baking of wheat bakery products. *Food Chem.* **2015**, *178*, 276–286. [CrossRef] [PubMed]

24. De Angelis, E.; Monaci, L.; Pascale, M.; Visconti, A. Fate of deoxynivalenol, T-2 and HT-2 toxins and their glucoside conjugates from flour to bread: An investigation by high-performance liquid chromatography high-resolution mass spectrometry. *Food Addit. Contam. Part A* **2013**, *30*, 345–355. [CrossRef] [PubMed]

25. Kostelanska, M.; Dzuman, Z.; Malachova, A.; Capouchova, I.; Prokinova, E.; Skerikova, A.; Hajslova, J. Effects of milling and baking technologies on levels of deoxynivalenol and its masked form deoxynivalenol-3-glucoside. *J. Agric. Food Chem.* **2011**, *59*, 9303–9312. [CrossRef] [PubMed]

26. Lancova, K.; Hajslova, J.; Kostelanska, M.; Kohoutkova, J.; Nedelnik, J.; Moravcova, H.; Vanova, M. Fate of trichothecene mycotoxins during the processing: Milling and baking. *Food Addit. Contam. Part A* **2008**, *25*, 650–659. [CrossRef] [PubMed]

27. Valle-Algarra, F.M.; Mateo, E.M.; Medina, Á.; Mateo, F.; Gimeno-Adelantado, J.V.; Jiménez, M. Changes in ochratoxin A and type B trichothecenes contained in wheat flour during dough fermentation and bread-baking. *Food Addit. Contam. Part A Chem. Anal. Control. Expo. Risk Assess.* **2009**, *26*, 896–906. [CrossRef] [PubMed]

28. Vidal, A.; Morales, H.; Sanchis, V.; Ramos, A.J.; Marín, S. Stability of DON and OTA during the breadmaking process and determination of process and performance criteria. *Food Control.* **2014**, *40*, 234–242. [CrossRef]

29. Wu, L.; Wang, B. Evaluation on levels and conversion profiles of DON, 3-ADON, and 15-ADON during bread making process. *Food Chem.* **2015**, *185*, 509–516. [CrossRef]

30. Schwartz-Zimmermann, H.E.; Hametner, C.; Nagl, V.; Fiby, I.; Macheiner, L.; Winkler, J.; Dänicke, S.; Clark, E.; Pestka, J.J.; Berthiller, F. Glucuronidation of deoxynivalenol (DON) by different animal species: Identification of iso-DON glucuronides and iso-deepoxy-DON glucuronides as novel DON metabolites in pigs, rats, mice, and cows. *Arch. Toxicol.* **2017**, *91*, 3857–3872. [CrossRef]
31. Roig, A.I.; Eskiocak, U.; Hight, S.K.; Kim, S.B.U.M.; Delgado, O.; Souza, R.F.; Spechler, S.J.; Wright, W.E.; Shay, J.W. Immortalized epithelial cells derived from human colon biopsies express stem cell markers and differentiate in vitro. *Gastroenterology* **2010**, *138*, 1012–1021. [CrossRef]

Article

Fate of Ergot Alkaloids during Laboratory Scale Durum Processing and Pasta Production

Sheryl A. Tittlemier *, Dainna Drul, Mike Roscoe, Dave Turnock, Dale Taylor and Bin Xiao Fu

Grain Research Laboratory, Canadian Grain Commission, 1404-303 Main Street, Winnipeg, MB R3C 3G8, Canada; dainna.drul@grainscanada.gc.ca (D.D.); mike.roscoe@grainscanada.gc.ca (M.R.); dave.turnock@grainscanada.gc.ca (D.T.); dale.taylor@grainscanada.gc.ca (D.T.); binxiao.fu@grainscanada.gc.ca (B.X.F.)

* Correspondence: sheryl.tittlemier@grainscanada.gc.ca; Tel.: +1-204-984-3456

Received: 6 March 2019; Accepted: 29 March 2019; Published: 31 March 2019

Abstract: The fate of ergot alkaloids during the milling of durum and subsequent production and cooking of pasta was examined. Durum samples containing varying amounts of ergot sclerotia (0.01–0.1% by mass) were milled, and all milling product was analyzed for 10 ergot alkaloids using liquid chromatography with tandem mass spectrometry. Spaghetti was prepared from the semolina obtained during milling. Ergocristine, ergocristinine, and ergotamine were the predominant ergot alkaloids observed in the milling fractions and spaghetti. Approximately 84% of the total ergot alkaloid mass of the whole grain durum resided in the milling product fractions associated with the outer kernel layers (bran, shorts, feeds). No consistent loss of ergot alkaloids was observed during the production or cooking of spaghetti. However, changes in the ratio of *R*- to *S*-enantiomers occurred during the milling and cooking of spaghetti. Products containing bran, shorts, and feeds, as well as cooked spaghetti, contained a higher proportion of the less biologically active *S*-enantiomers. The results of this study emphasize the need to monitor *R*- and *S*-enantiomers, and to consider food and feed products, as opposed to whole grain, when assessing any exposure of consumers to ergot alkaloids.

Keywords: mycotoxin; milling; bran; semolina; cooking; dietary exposure

Key Contribution: No substantial losses of ergot alkaloids were associated with boiling pasta. However, cooking pasta resulted in epimerization and a shift towards higher concentrations of the less biologically active *S*-enantiomers.

---

1. Introduction

Ergot alkaloids (EAs) are a group of mycotoxins produced during the fungal infection of cereals. In Canada, the fungus *Claviceps purpurea* causes ergot infection of rye, barley, oats, wheat, and durum [1]. During *C. purpurea* infection, healthy kernels are replaced by dark-colored sclerotia that contain high concentrations of various EAs [2–4]. The incidence of ergot infection of durum grown in Canada is variable amongst growing years, but appears to be increasing since the early 2000s [5]. This increase suggests that the potential for EAs occurring in durum, and durum-based food products, is also increasing. A number of EAs have been measured in Canadian cereals [5], as well as those grown in western [6] and eastern Europe [7], and Australia [3].

The EAs most commonly associated with cereal grains are amide-like derivatives of lysergic acid [8]. These compounds contain a stereogenic center, and can exist in *R*- (indicated by an "ine" suffix in their names) and *S*- (indicated by an "inine" suffix in their names) enantiomeric forms. The enantiomers can undergo reversible epimerization, which can reach equilibrium. Epimerization has been reported to occur in solution under alkaline conditions and lengthy periods of storage [9].

Consumption of ergot alkaloids has been long known to cause ergotism in both humans and livestock [10,11]. Symptoms of ergotism include gangrene, gastrointestinal effects, reduced lactation, as well as effects on the central nervous system.

Current practices to avoid or minimize exposure to ergot alkaloids in food and feed involve grain handling and milling procedures to exclude ergot sclerotia from entering food and feed channels. Tolerances are also used to regulate the amount of ergot sclerotia in grain. For example, the Canadian maximum level for ergot sclerotia in durum is 0.02% on a mass basis for the higher quality grades of No.1 Canada Western Amber Durum (CWAD) and No.2 CWAD [12], and the Codex Alimentarius Commission maximum level of ergot sclerotia in wheat is 0.05% [13].

Currently, no jurisdictions have set maximum levels (MLs) for ergot alkaloids in grain, food, or feed. However, regulatory agencies are exploring the use of MLs as a tool to manage health risk from exposure to EAs. The Canadian Food Inspection Agency (CFIA) consulted stakeholders on proposed limits for a number of contaminants, including ergot alkaloids, in livestock feeds [14]. The complexity of EA exposure and hazard assessments due to processing effects on EA concentrations, toxicity of individual EAs, and synergistic or antagonistic effects in EA mixtures, as well as variable susceptibilities due to age, sex, and physiological state, coupled with limited published scientific information in these areas, led the CFIA to revisit setting MLs for EAs in the future [15].

There are limited published studies on the fate of ergot alkaloids in food or feed products during the processing of grain. Most of the studies focus on rye-based food and feed. Fajardo et al. [16] investigated the fate of six EAs in milling products of red spring wheat. Flour containing EAs was also used to prepare pasta, Asian noodles, and bread. Overall, EAs were concentrated in the milling products associated with reduction streams, and some losses were noted in cooked noodles. However, a limitation of this work is the inclusion of only *R*-enantiomers in the EA analyses. Merkel et al. [17] studied the fate of twelve EAs in cookies baked with rye flour and subjected to in vitro digestion. The authors reported degradation of EAs during baking, as well as a shift from *R*- to *S*-enantiomers for all EAs. A feeding study conducted by Dänicke [18] included an assessment of the fate of twelve EAs in rye-based feed for laying hens, which had been processed under heat and pressure. Substantial epimerization of EAs was noted after processing, with an increase in concentration of *S*-enantiomers for all EAs.

The objective of this work was to take advantage of ongoing pasta quality and functionality studies, and examine the fate of EAs in milled durum, as well as during the processing and cooking of spaghetti. This study used durum, a cereal grain important in the production of pasta. The fate of EAs during the processing of durum has not been well studied. This study also monitored the fate of four *R/S* enantiomeric pairs, as well as two additional EA *R*-enantiomers for which the *S*-enantiomer was not readily available at the time of the study. The outcomes of this study are valuable for use in exposure assessments to ensure that consumers' exposure to EAs in food, or feed, is accurately determined.

2. Results

2.1. Ergot Alkaloids in Whole Grain and Milling Products

Table 1 lists the EA concentrations measured in the comminuted whole grain and milling products for the six samples of CWAD containing varying amounts of ergot sclerotia. Concentrations are provided as the sum of 10 ergot alkaloid analyte concentrations. Milling yields from the two mill runs of 2 kg each are also listed in Table 1.

Total EA concentrations for whole grain and all milling products increased as the percentage of ergot sclerotia in the original durum sample increased. The correlation between the percentage of ergot sclerotia and total EA concentration was statistically significant for whole grain and all milling products (Pearson Product Moment Correlation, $p < 0.0025$), aside from bran ($p < 0.111$). The total EA concentration of bran increased from 1112 to 5695 µg/kg as the ergot sclerotia content increased from 0.01% to 0.1%, and reached a maximum of 5715 µg/kg at an ergot sclerotia content of 0.04%.

Table 1. Total ergot alkaloids concentration (µg/kg) in whole grain durum and milling products. Average ± standard deviation (2 × 2 kg mill runs for each of the 6 samples) milling yields are provided for the milling products.

	Whole Grain	Bran	Shorts	Feeds	Semolina 1	Semolina 2	Semolina 3	Flour	3rd Break Flour
Milling yield (%, m/m)		12.4 ± 0.1	5.5 ± 0.1	5.3 ± 0.1	61.3 ± 0.1	5.2 ± 0.1	2.39 ± 0.04	3.49 ± 0.07	4.46 ± 0.05
No ERG	<2	<2	<2	12	16	<2	<2	<2	<2
ERG 0.01%	446	1112	241	946	23	337	406	177	188
ERG 0.02%	601	1392	2267	2231	173	368	836	376	313
ERG 0.03%	681	2996	2298	3173	186	461	1005	506	412
ERG 0.04%	936	5715	3482	4532	279	1358	1300	618	530
ERG 0.1%	2147	5695	8002	9955	579	1603	3750	1493	1165

EAs were observed in all fractions from the five samples that contained ergot sclerotia. Only feeds and the first semolina fraction from the CWAD sample with no visible ergot sclerotia contained measurable EAs. Total EA concentrations in these fractions were low, ranging from 12 to 16 µg/kg. Total EA concentrations from the other samples were highest in bran, shorts, and feeds and the lowest was in the first semolina fraction. The concentrations in this semolina fraction were approximately 20–40× lower than the maximum measured in bran, shorts, or feeds.

Ergocristine and ergocristinine, followed by ergotamine, were the most predominant EAs observed in whole grain and milling products. The mean percentage of total EAs on a molar (as opposed to mass-based concentration) basis for each EA in the five samples containing ergot sclerotia was determined; these are presented in Figure 1.

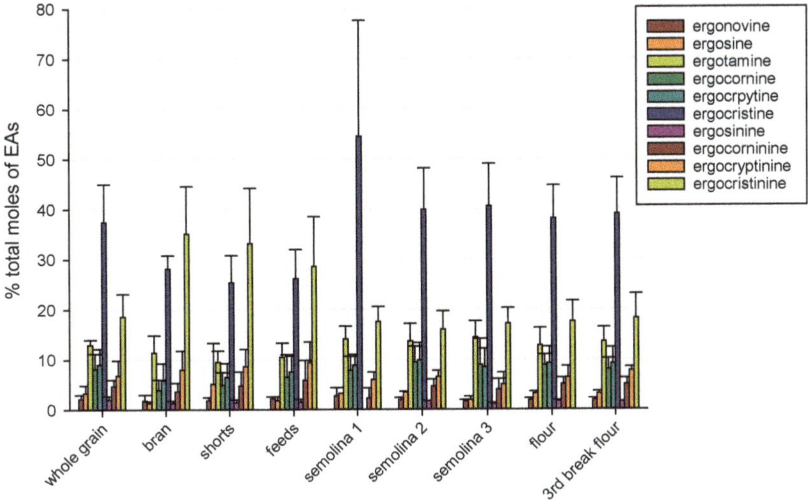

Figure 1. Mean ± standard deviation percentage of total moles of ergot alkaloids present in whole grain and milling products. Means and standard deviations were calculated from whole grain and milling products from the five durum samples with ergot sclerotia ranging from 0.01–0.1% by mass.

In bran, shorts, and feeds, ergocristinine was the predominant EA at 29–35% of total EAs. The molar percentage of ergocristinine in bran and shorts were significantly different from those of the semolina and flour fractions ($p < 0.05$). Ergocristine was predominant in whole grain and other milling products, ranging from 37–55% of total EAs. The mean molar percentage of ergocristine was significantly different between semolina 1 and bran, shorts, and feeds ($p < 0.05$).

No other significant differences were noted amongst the other milling products. Ergotamine ranged from 10–14% of total EAs in whole grain and all milling products.

2.2. Ergot Alkaloids in Spaghetti

Table 2 lists the mean recoveries of individual EAs from fortified boiled spaghetti and cooking water. Mean recoveries for all EAs ranged from 93% to 113% from cooking water. Mean recoveries from boiled spaghetti were assessed only for the *R* enantiomers in order to preserve standard material, and because no difference in recoveries was observed for cooking water and whole grain [5]. Mean recoveries of the *R* enantiomer EAs from fortified boiled spaghetti ranged from 98% to 143%.

Table 2. Mean ± standard deviation percent recovery of ergot alkaloids from cooking water and cooked spaghetti. Cooking water samples were fortified with ergot alkaloids to produce a concentration of 40 µg/kg for ergonovine, 200 µg/kg for the other *R*-enantiomers and 100 µg/kg for the *S*-enantiomers. Boiled spaghetti samples were fortified to produce a concentration of 40 µg/kg for ergonovine and 200 µg/kg for the other *R*-enantiomers; spaghetti was not fortified with *S*-enantiomers. Triplicate replicates of cooking water and boiled spaghetti were analyzed for *R*-isomers. Duplicate replicates of cooking water were analyzed for *S*-isomers.

	Cooking Water	Boiled Spaghetti
Ergonovine	102 ± 3	143 ± 5
Ergosine	97 ± 5	102 ± 1
Ergotamine	100 ± 1	102 ± 1
Ergocornine	103 ± 1	98 ± 1
Ergocryptine	103 ± 4	100 ± 1
Ergocristine	113 ± 2	143 ± 5
Ergosinine	98 ± 9	-
Ergocorninine	93 ± 7	-
Ergocryptinine	104 ± 9	-
Ergocristinine	112 ± 11	-

Table 3 lists concentrations measured in freshly extruded spaghetti, cooked spaghetti, and cooking water from the six samples of CWAD containing varying amounts of ergot sclerotia. The moisture contents of the spaghetti are also provided. The concentrations provided in Table 3 are on a fresh weight basis; that is, they are not normalized to a specific moisture content. The moisture content of cooked spaghetti was approximately twice that of the freshly extracted spaghetti.

Table 3. Mean ± standard deviation total ergot alkaloids concentration (µg/kg) in freshly extruded spaghetti, cooked spaghetti, and remaining cooking water.

	Freshly Extruded Spaghetti [1]	Cooked Spaghetti [2]	Cooking Water [2]
Moisture content (%, m/m)	34	66	-
No ERG	<2	<2	<2
ERG 0.01%	42.2 ± 0.9	26	<2
ERG 0.02%	114 ± 12	101	2
ERG 0.03%	143 ± 13	84	2
ERG 0.04%	175 ± 13	124	5
ERG 0.1%	533 ± 13	244	10

[1] $n = 2$; [2] $n = 1$.

Similar to whole grain, semolina, and flour fractions, ergocristine, ergocristinine, and ergotamine were the most predominant EAs observed in freshly extruded spaghetti. The mean percentage and standard deviation of total EAs on a molar basis for these three EAs in the five samples containing ergot sclerotia were 42 ± 8% for ergocristine, 19 ± 3% for ergocristinine, and 15 ± 5% for ergotamine. For cooked spaghetti and cooking water, ergocristinine was the predominant EA at 55 ± 8% and 73 ± 28% of the total EA molar content, respectively.

3. Discussion

3.1. Fate of Ergot Alkaloids in Milling Products and Spaghetti

Most of the EAs associated with sclerotia were in the by-products of semolina milling, i.e., bran, shorts, and feeds from outer kernel layers. Higher amounts of EAs were associated with bran, shorts, feeds (Figure 2), which are milling products containing material from the outer kernel. These three milling products contained 84% of the total EA mass in the durum.

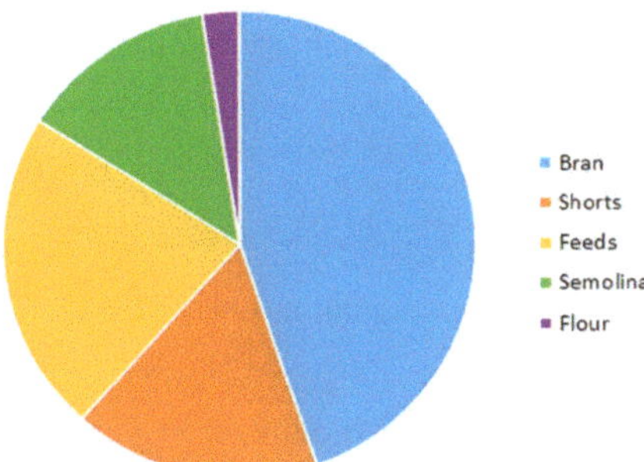

Figure 2. Distribution of total ergot alkaloids mass amongst durum milling products. Pie slices represent the mean fraction of total ergot alkaloid mass across the five durum samples with ergot sclerotia ranging from 0.01–0.1% by mass.

The presence of the EAs predominantly in these milling fractions is consistent with the observation by Franzmann et al. [19] that the amounts of EAs in rye flour increased with a higher amount of bran in the flour. Franzmann et al.'s work attributed this EA distribution on rye kernels to the coincidental contact and abrasion between sound rye and ergot sclerotia during ordinary grain handling and movement.

Even though EAs will be transferred to sound kernels from ergot sclerotia during movement, the fate of ergot sclerotia during the milling process will be the most important factor affecting EA content of milling products because the concentration of EAs in ergot sclerotia are orders of magnitude greater than in sound grain. Total EA concentrations in rye ergot sclerotia were approximately $300\times$ greater than concentrations in rye that had been mixed with ground ergot sclerotia and subject to cleaning [19]. Mean concentrations of the 10 EAs included in the current study ranged from 500 to 1000 mg/kg in ergot sclerotia obtained from infected durum plants [4].

The predominance of EAs in the bran, shorts, and feeds milling products fractions is also consistent with the path taken by ergot sclerotia through a wheat milling procedure reported by Farjado et al. [16]. The aforementioned Farjado et al. [16] milled wheat containing various amounts of ergot sclerotia. They noted that the sclerotia congregated in the reduction system because they did not flake when passing through the break rolls. This led to approximately 75% of EAs present in the whole grain wheat residing in bran and shorts milling fractions in their study.

The fate of the individual EAs was also examined over the milling process. Interestingly, there were differences in the EA profiles amongst the milling products associated with the outer kernel layers (bran, shorts, and feeds) and endosperm (semolina, flour), in addition to the variation in concentration.

These differences in EA profiles are illustrated in Figure 3. The ratio of *R*- to *S*-enantiomers was lower in milling products associated with the outer kernel layers as compared to whole grain, indicating a predominance of the *S*-enantiomers. The *R*-enantiomers were predominant in semolina and flour, and were present in these fractions to greater extent than in whole grain.

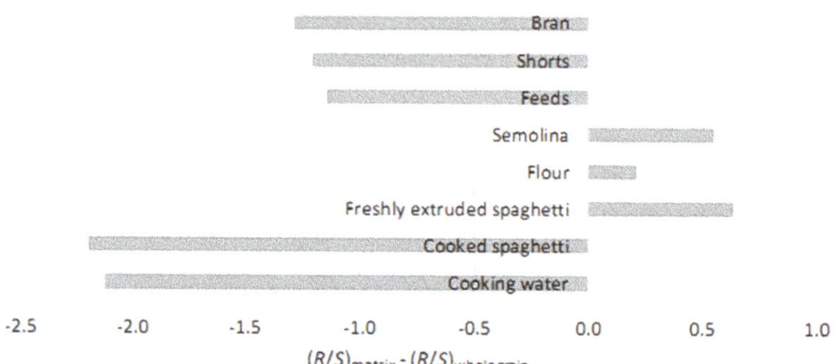

$(R/S)_{matrix} - (R/S)_{wholegrain}$

Figure 3. Difference in ratios of R-enantiomer to S-enantiomer concentrations between whole grain durum, milling products, and pasta matrices. The difference in ratios was calculated using mean R/S concentration ratios determined from the five groups containing 0.01–0.1% ergot sclerotia by mass.

The work in this study does not directly address or examine the cause of the differences in occurrence amongst individual EAs. However, past research has investigated or noted the epimerization of EAs. Epimerization of EAs is reported to be promoted by exposure to light [20], therefore EAs in the outer kernel layers may be subject to more light and subsequent epimerization than EAs in the inner kernel layers. Even though heat also appears to facilitate epimerization [17,18], it is unlikely that the milling process used in this study promoted epimerization in the bran, shorts, and feeds fractions, as temperatures generated during the roller milling of wheat are around 35 °C [21].

The cooking of spaghetti did not appear to considerably affect the presence of EAs. No substantial losses of EAs were consistently evident after the preparation of spaghetti by extrusion, nor after boiling the spaghetti. Figure 4 provides a comparison of the amount of EAs in freshly extruded spaghetti, cooked spaghetti, and the cooking water, to semolina. The comparison is on the basis of EA mass, therefore the impact from varying moisture content of the products is avoided. Across the five durum samples with varying ergot sclerotia contents, freshly extruded spaghetti and cooked spaghetti samples contained 77 ± 14% and 93 ± 18% of the total EA content observed in semolina, respectively. Cooking water contained a negligible amount of EAs (0.04 ± 0.02%) of the total semolina EA content.

Overall, the results observed do not indicate a consistent and extensive loss of EAs during processing. While the amounts of EAs observed in the freshly extruded spaghetti seem to suggest some loss, the amounts of EAs observed in cooked spaghetti do not demonstrate similar losses. The apparently lower EA amounts measured in the freshly extruded spaghetti may reflect differences in the ability of the analytical method to extract and/or measure EAs in this matrix.

Merkel et al. [17] reported small losses of 2–30% in cookies due to degradation of EAs during baking. Dänicke noted an average loss of 11% for EAs in heat treated rye [18], but the changes in concentrations across the five chicken diets examined, ranged from a loss of 26%, to an increase of 15%. This inconsistency suggests that the heterogeneous nature of ergot contamination of whole grain may have contributed to the apparent loss of EAs in the heat treated rye.

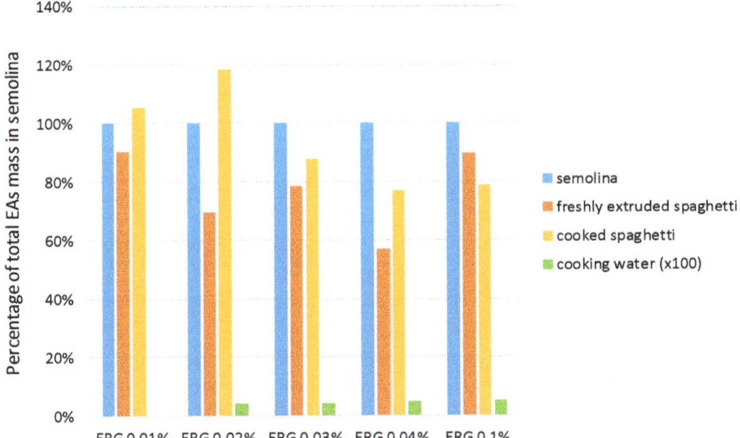

Figure 4. Amount of total ergot alkaloids in processed spaghetti and cooking water relative to semolina. The mean of duplicates is reported for freshly extruded spaghetti. Single analyses were performed for the other matrices.

Fajardo et al. [16] also reported losses of EAs during the cooking of Asian noodles and spaghetti made with wheat flour. However the analytical method used in that study only included the *R*-enantiomers as analytes. Any occurrence of *S*-enantiomers formed by epimerization during cooking would not be observed. Therefore changes in concentrations of EAs due to epimerization would appear instead as losses of EAs.

Even though no substantial losses of EAs were observed during the processing and cooking of spaghetti, the EA profile did change during cooking. As seen in Figure 3, the EA content of semolina and freshly extruded spaghetti is dominated by *R*-enantiomers, which changed to a predominance of *S*-enantiomers in cooked spaghetti and cooking water. The changes in the EA profile observed suggest that the heat of extrusion (45 °C) used to prepare the fresh spaghetti is not enough to promote epimerization, whereas the heat of boiling water can facilitate epimerization.

The enrichment of *S*-enantiomers after observed after cooking spaghetti is consistent with changes observed in other research. Merkel et al. reported the epimeric ratio shifted toward the *S*-enantiomer for all EAs in baked cookies [17]. Dänicke [18] also noted a consistent increase in the proportion of *S*-enantiomers (and concomitant decrease in *R*-enantiomers) in rye with varying ergot sclerotia content after heat treatment.

3.2. Implications of Ergot Alkaloid Fate in Milling Products and Spaghetti

The wider implications of this work relate to the distribution of EAs amongst the durum milling products and the epimerization of EAs observed. The association of EAs with the bran, shorts, and feeds fractions after durum milling will lower the exposure for populations consuming food products, such as pasta, made from semolina, as compared to whole grain durum. In turn, any incorporation of bran, shorts, and feeds fractions into animal feed will increase exposure of livestock, as compared to use of whole grain durum.

The epimerization observed from cooking and milling durum indicate that feed products and cooked pasta will contain a higher proportion of the less biologically active *S*-enantiomers. Overall, the deleterious health effects of EAs are associated with the ability of EAs to act as ligands for a variety of receptors. The ligand activity is reported to be greater for *R*-enantiomers, compared to *S*-enantiomers [18,20].

The fractionation of EAs amongst milling products and the epimerization observed during milling and cooking highlights the need for exposure assessments to consider concentrations in food products, or to use a processing factor, as opposed to using whole grain durum EA concentrations to estimate consumers' exposure. The results of this study also indicate that both *R*- and *S*-enantiomers should be monitored and assessed, in order to obtain an accurate view of consumers' exposure to EAs.

4. Materials and Methods

4.1. Samples

A No.1 Canada Western Amber Durum sample (CWAD, 30 kg) was cleaned and hand-picked to remove ergot bodies and any kernels with dark discoloration. The cleaned CWAD wheat that did not contain ergot sclerotia was divided into 6 × 5 kg sub-samples. Sclerotia hand-picked out of other naturally-infected CWAD samples were added back to five of the sub-samples in order to prepare grain with 0.01, 0.02, 0.03, 0.04, and 0.1% (m/m) ergot sclerotia. The sixth 5 kg sub-sample was kept at 0% ergot sclerotia. The 5 kg sub-samples were divided into 2 × 2 kg and 1 × 1 kg portions using a rotary sample divider (Materials Sampling Solutions, Southport, Australia). The 2 × 2 kg portions were milled as described below. The 1 kg whole grain portion was comminuted using a Retsch SR 300 rotor beater mill fitted with a 750 μm screen and coupled with a Retsch DR 100 vibratory feeder.

4.2. Milling and Processing into Spaghetti

Durum wheat samples were milled on a four stand Allis-Chalmers laboratory mill coupled with a small-scale semolina purifier previously described by Dexter et al. [22]. The mill flow consists of four corrugated break roll passages, five corrugated sizing roll passages, and 10 purification steps. The mill room was controlled at 21 °C and 60% relative humidity. Durum wheat samples were tempered to 16% moisture for 16 h before milling. Particles retained on a 425 μm screen after the four break passages were collected as bran. After each break passage, coarse material retained on a 630 μm screen was passed through sizing rolls. After sizing passages, flour was collected by combining materials with fine particles passing through the 180 μm sieve, and particles retained on a 700 μm screen were collected as shorts. The remaining milling streams were passed through a purifier. After each purification step, streams with particles passing through the 571 μm screens, but retained on the 183 μm screens, were collected as semolina. Particles retained on the 630 μm screens on purifiers 7–10 were collected as feed.

Following the method of Fu et al. [23], spaghetti was produced from semolina using a customized micro-extruder (Randcastle Extrusion Systems Inc., Cedar Grove, NJ, USA). For dough preparation, semolina was mixed with water in a high speed asymmetric centrifugal mixer (DAC 400 FVZ SpeedMixer, FlackTec, Landum, SC, USA) at constant water absorption of 31.5%. Vacuum was applied to eliminate the introduction of air bubbles, after which the dough crumbs were extruded through a four-hole Teflon coated spaghetti die (1.8 mm).

The fresh pasta was subsequently dried in a pilot pasta dryer (Bühler, Uzwil, Switzerland) coupled with a 325 min drying cycle and maximum temperature of 85 °C.

4.3. Cooking and Moisture Content Determination

Spaghetti (10 g) was added to 100 g of water brought to boil. The spaghetti was cooked uncovered for 10 min. After 10 min, cooked spaghetti was immediately removed from the cooking water with tongs, excess adhered water was left to evaporate for 1 min, and the mass of cooked spaghetti was determined. A 10 g portion of boiled spaghetti was then placed into a polyethylene sample bottle, which was closed and stored for approximately 2 h until extraction and EA analysis. The remainder of the boiled spaghetti was re-weighed, stored in sealed plastic bags, and frozen until moisture content determination. The remaining cooking water was cooled in a refrigerator and stored for approximately 4 h until extraction and EA analysis.

The moisture content of cooked spaghetti was determined gravimetrically. The remainders of the boiled spaghetti that were weighed immediately after cooking were placed on sieves and air dried for 24 h. The air dried samples were ground using a Brabender MLI-204 break mill, and then dried for 1 h in an oven held at 130 °C. The moisture content of the cooked spaghetti was calculated from the decrease between the post-cooking and air dried masses as well as the pre- and post-oven dried masses.

Moisture content of freshly extruded spaghetti was calculated based on the amount of water added and originally present in the semolina, minus 0.5% during to moisture loss during pasta processing.

4.4. Determination of Ergot Alkaloids

Samples were analyzed for 10 ergot alkaloids according to the liquid chromatography tandem mass spectrometry method described by Tittlemier et al. [5]. The 10 ergot alkaloid analytes are listed in Table 2. Four of the alkaloids were *S*-epimers of *R*-enantiomers. The two *S*-epimers that were not included in the method (ergonovinine and ergotaminine) were not easily obtained at the time of analysis. In order to correct for variations in final extract volume and injection volume, dihydroergotamine was used as an internal standard.

Milling products and spaghetti (10 g) were extracted with 50 mL 84:16 (*v/v*) acetonitrile/3.03 mM aqueous ammonium carbonate. The slurries of milling products were shaken on a flatbed shaker for 30 min, and spaghetti was comminuted in extraction solvent using a handheld laboratory homogenizer for 3 min at 12,000 rpm. After extraction, sample extracts were then centrifuged, and an aliquot of supernatant was diluted with 3.03 mM aqueous ammonium carbonate. Internal standard was added to all samples prior to analysis. Post-extraction fortified calibration standards were prepared in wheat matrix.

Water (1 g) was transferred from the cooled cooking water, combined with 1 mL acetonitrile, and vortex mixed for 10 s. The solution was left to sit for 2 min prior to a 1 mL aliquot being taken and combined with 1.5 mL of 3.03 mM aqueous ammonium carbonate. Internal standard was added, final sample extracts were then vortex mixed and filtered using PTFE filters prior to analysis.

Sample extracts were chromatographed on a C_{18} column followed by analysis in the positive electrospray ionization mode using multiple reaction monitoring. Two transitions were monitored for each analyte. In order to avoid epimerization of ergot alkaloids, the exposure of sample extracts to light was minimized by covering samples and extracts during processing, and by keeping the illumination option off on the liquid chromatograph sample manager.

Analytes were considered to be positively identified and quantitated if their retention times were within 0.1 min of the average retention time of the corresponding analyte in the external calibration standards; the peak had a signal-to-noise ratio greater than 9:1, and the ratio of qualification to quantitation ions was within acceptable tolerances [24]. Analyte peak areas were normalized to the dihydroergotamine peak area in samples during calculation of concentrations.

Blank wheat samples fortified with a solution of standards and an in-house rye reference material were analyzed with each batch of samples and were used to monitor the performance of the analytical method during the study.

4.5. Evaluation of Method for Analysis of Cooking Water and Boiled Spaghetti

Because water and boiled spaghetti were different matrices than the grain used in the initial validation of the EA analytical method [5], additional method evaluation was performed for these two matrices. Boiled commercially-available durum spaghetti, and the cooking water used to boil the spaghetti, were fortified with ergot alkaloids, and analyzed as described above.

The boiled spaghetti was fortified with the *R*-enantiomers only; fortification concentrations were 200 µg/kg for each alkaloid aside from ergonovine, which was fortified at 40 µg/kg. The boiled spaghetti was drained, and as much water adhering to the cooked pasta as possible was removed by shaking. The cooked spaghetti was then weighed within a minute after boiling to obtain a relevant cooked mass, and 10 g was transferred to a 250 mL centrifuge bottle, capped, and stored in darkness for 30 min to cool prior to fortification. After fortification, samples sat for 20 min in darkness at room temperature prior to extraction. Extraction and analysis was performed as described above, using a ratio of 5:1 (v/m) extraction solvent to mass of boiled spaghetti.

The cooking water was fortified with both *R*- and *S*-enantiomers. After the cooked spaghetti was removed, the water was placed in a refrigerator for 30 min. After cooling, two portions of 10 mL were removed. One portion was fortified with *R*-enantiomers, and the other was fortified with *S*-enantiomers. Concentrations of ergot alkaloids in both portions were as mentioned above for boiled spaghetti. Three aliquots were taken from the *R*-enantiomers fortified 10 mL, and two aliquots were taken from the *S*-enantiomers fortified 10 mL. Extraction and analysis was performed as described above. All aliquots were analyzed for all ergot alkaloids in order to examine if epimerization had occurred during sample processing.

4.6. Statistical Analyses

Statistical analyses were performed using SigmaPlot 13.0 (Systat Software Inc., Chicago, IL, USA). The mean molar percentages of ergocristine and ergocristinine were compared amongst milling products using a Kruskal-Wallis One Way Analysis of Variance on Ranks. The Tukey Test was used to isolate milling products whose mean molar percentages differed from others. Statistical tests were limited to these two EAs because they were the predominant compounds in the milling products.

Author Contributions: Conceptualization, S.A.T.; Methodology, B.X.F and S.A.T.; Formal Analysis, S.A.T.; Investigation, D.D., M.R., D.T. (Dave Turnock), and D.T. (Dale Taylor); Resources, B.X.F. and S.A.T.; Writing-Original Draft Preparation, S.A.T.; Supervision, B.X.F. and S.A.T.; Project Administration, S.A.T.

Funding: This research received no external funding.

Conflicts of Interest: The authors declare no conflict of interest.

References

1. Menzies, J.G.; Turkington, T.K. An overview of the ergot (*Claviceps purpurea*) issue in western Canada: Challenges and solutions. *Can. J. Plant Pathol.* **2015**, *37*, 40–51. [CrossRef]
2. Appelt, M.; Ellner, F.M. Investigations into the occurrence of alkaloids in ergot and single sclerotia from the 2007 and 2008 harvests. *Mycotoxin Res.* **2009**, *25*, 95–101. [CrossRef] [PubMed]
3. Blaney, B.J.; Molloy, J.B.; Brock, I.J. Alkaloids in australian rye ergot (*Claviceps purpurea*) sclerotia: Implications for food and stockfeed regulations. *Animal Product. Sci.* **2009**, *49*, 975–982. [CrossRef]
4. Tittlemier, S.A.; Drul, D.; Roscoe, M.; Menzies, J.G. The effects of selected factors on measured ergot alkaloid content in *Claviceps purpurea*-infected hexaploid and durum wheat. *World Mycotoxin J.* **2016**, *9*, 555–564. [CrossRef]
5. Tittlemier, S.A.; Drul, D.; Roscoe, M.; McKendry, T. Occurrence of ergot and ergot alkaloids in western Canadian wheat and other cereals. *J. Agricult. Food Chem.* **2015**, *63*, 6644–6650. [CrossRef] [PubMed]
6. Orlando, B.; Maumené, C.; Piraux, F. Ergot and ergot alkaloids in French cereals: Occurrence, pattern and agronomic practices for managing the risk. *World Mycotoxin J.* **2017**, *10*, 327–338. [CrossRef]
7. Topi, D.; Jakovac-Strajn, B.; Pavšič-Vrtač, K.; Tavčar-Kalcher, G. Occurrence of ergot alkaloids in wheat from Albania. *Food Addit. Contamin A* **2017**, *34*, 1333–1343. [CrossRef] [PubMed]
8. Stoll, A.; Hoffmann, A. The chemistry of the ergot alkaloids. In *Chemistry of the Alkaloids*; Pelletier, S.W., Ed.; Van Nostrand Reinhold Company: New York, NY, USA, 1970; pp. 267–301.
9. Krska, R.; Stubbings, G.; Macarthur, R.; Crews, C. Simultaneous determination of six major ergot alkaloids and their epimers in cereals and foodstuffs by LC–MS–MS. *Anal. Bioanal. Chem.* **2008**, *391*, 563–576. [CrossRef] [PubMed]
10. Richard, J.L. Some major mycotoxins and their mycotoxicoses—An overview. *Int. J. Food Microbiol.* **2007**, *119*, 3–10. [CrossRef] [PubMed]
11. Belser-Ehrlich, S.; Harper, A.; Hussey, J.; Hallock, R. Human and cattle ergotism since 1900: Symptoms, outbreaks, and regulations. *Toxicol. Ind. Health* **2012**, *29*, 307–316. [CrossRef] [PubMed]
12. Canadian Grain Commission. Official Grain Grading Guide. Available online: https://www.grainscanada.gc.ca/oggg-gocg/ggg-gcg-eng.htm (accessed on 27 February 2019).

13. Codex Alimentarius Commission. Codex Standard for Wheat and Durum Wheat. Codex Stan 199-1995. Available online: http://www.fao.org/fao-who-codexalimentarius/sh-proxy/en/?lnk=1&url=https%253A%252F%252Fworkspace.fao.org%252Fsites%252Fcodex%252FStandards%252FCODEX%2BSTAN%2B199-1995%252FCXS_199e.pdf (accessed on 27 February 2019).
14. Canadian Food Inspection Agency. Proposal-Contaminant Standards for Aflatoxins, Deoxynivalenol, Fumonisins, Ergot Alkaloids and Salmonella in Livestock Feeds. Available online: http://www.inspection.gc.ca/animals/feeds/consultations/contaminant-standards-for-aflatoxins-deoxynivaleno/eng/1500908795245/1500908795965 (accessed on 27 February 2019).
15. Canadian Food Inspection Agency. What We Heard Report-Consultations on Contaminant Standards for Aflatoxins, Deoxynivalenol, Fumonisins, Ergot Alkaloids and Salmonella in Livestock Feeds. Respondent Comments and Cfia Responses. Available online: http://www.inspection.gc.ca/animals/feeds/consultations/contaminant-standards-for-aflatoxins-deoxynivaleno/consultation-summary/eng/1544482075311/1544482140088 (accessed on 27 February 2019).
16. Fajardo, J.E.; Dexter, J.E.; Roscoe, M.M.; Nowicki, T.W. Retention of ergot alkaloids in wheat during processing. *Cereal Chem.* **1995**, *72*, 291–298.
17. Merkel, S.; Dib, B.; Maul, R.; Köppen, R.; Koch, M.; Nehls, I. Degradation and epimerization of ergot alkaloids after baking and in vitro digestion. *Anal. Bioanal. Chem.* **2012**, *404*, 2489–2497. [CrossRef] [PubMed]
18. Dänicke, S. Toxic effects, metabolism, and carry-over of ergot alkaloids in laying hens, with a special focus on changes of the alkaloid isomeric ratio in feed caused by hydrothermal treatment. *Mycotoxin Res.* **2016**, *32*, 37–52. [CrossRef] [PubMed]
19. Franzmann, C.; Schröder, J.; Münzing, K.; Wolf, K.; Lindhauer, M.; Humpf, H.-U. Distribution of ergot alkaloids and ricinoleic acid in different milling fractions. *Mycotoxin Res.* **2011**, *27*, 13–21. [CrossRef] [PubMed]
20. European Food Safety Authority. Scientific opinion on ergot alkaloids in food and feed. *EFSA J.* **2012**, *10*, 2798–2956.
21. Prabhasankar, P.; Haridas Rao, P. Effect of different milling methods on chemical composition of whole wheat flour. *Eur. Food Res. Technol.* **2001**, *213*, 465–469. [CrossRef]
22. Dexter, J.E.; Matsuo, R.R.; Kruger, J. The spaghetti-making quality of commercial durum wheat samples with variable α-amylase activity. *Cereal Chem.* **1990**, *67*, 405–412.
23. Fu, B.; Schlicting, L.; Pozniak, C.; Singh, A. Pigment loss from semolina to dough: Rapid measurement and relationship with pasta colour. *J. Cereal Sci.* **2013**, *57*, 560–566. [CrossRef]
24. European Commission. Commission decision implementing council directive 96/23/EC concerning the performance of analytical methods and the interpretation of results. *Off. J. Eur. Com.* **2002**, *221*, 8–36.

Review

Mycotoxins during the Processes of Nixtamalization and Tortilla Production

Sara Schaarschmidt * and Carsten Fauhl-Hassek

German Federal Institute for Risk Assessment (BfR), Department Safety in the Food Chain, Max-Dohrn-Str. 8-10, D-10589 Berlin, Germany; carsten.fauhl-hassek@bfr.bund.de
* Correspondence: sara.schaarschmidt@bfr.bund.de

Received: 28 March 2019; Accepted: 11 April 2019; Published: 16 April 2019

Abstract: Tortillas are a traditional staple food in Mesoamerican cuisine, which have also become popular on a global level, e.g., for wraps or as snacks (tortilla chips). Traditional tortilla production includes alkaline cooking (nixtamalization) of maize kernels. This article summarizes the current knowledge on mycotoxin changes during the nixtamalization of maize and tortilla production. Upon nixtamalization, mycotoxins can be affected in different ways. On the one hand, the toxins can be physically removed during steeping and washing. On the other hand, mycotoxins might be degraded, modified, or released/bound in the matrix by high pH and/or high temperature. This also applies to the subsequent baking of tortillas. Many studies have shown reduced mycotoxin levels in alkali-cooked maize and in tortillas. Most of the available data relate to aflatoxins and fumonisins. The reduction (and detoxification) of aflatoxins during nixtamalization might, however, be partially reversed in acidic conditions. The loss of fumonisin concentrations is to some extent accompanied by hydrolyzation and by lower toxicity. However, some studies have indicated the potential formation of toxicologically relevant modified forms and matrix-associated fumonisins. More data are required to assess the influence of alkaline cooking regarding such modified forms, as well as mycotoxins other than aflatoxins/fumonisins.

Keywords: aflatoxins; alkaline; hydrolyzed fumonisins; fumonisins; food processing; maize; masa; matrix-associated mycotoxins; modified mycotoxins; tortillas

Key Contribution: The paper provides a critical overview of the effect of masa and tortilla production on mycotoxin concentrations considering the potential degradation and transformation of mycotoxins and matrix–toxin interactions.

1. Introduction

Mycotoxins are secondary fungal metabolites that are produced in the field and/or during the storage of crops and raise health concerns for humans and animals due to their toxic potential. Typically, several mycotoxins occur in parallel in crops. They can be produced by different fungal species, but single species are also usually capable of producing a distinct set of toxins [1]. Aflatoxins are mainly produced by *Aspergillus* and *Pencillium* species, with aflatoxin B1 (AFB1) being the most toxic and carcinogenic. Aflatoxins—including AFB1, aflatoxin B2 (AFB2), aflatoxin G1 (AFG1), and aflatoxin G2 (AFG2)—are particularly common in maize and other crops produced in warmer climates and are a serious health threat in many regions worldwide (for more information see e.g., [2,3]). Fumonisins belong to a large group of toxins referred to as *Fusarium* toxins that are produced by several *Fusarium* species, such as *Fusarium verticillioidies*. Moreover, fumonisins can be produced by some species of the *Aspergillus niger* complex. Fumonisins B1, B2, and B3 (FB1, FB2, FB3) are frequently found in raw maize and can exhibit liver and kidney toxicity [4]. Other *Fusarium* toxins, which are often present in maize, include zearalenone (ZEN [5]) and trichothecenes, such as deoxynivalenol (DON) [6,7].

In addition to such 'traditional' mycotoxins, which have been assessed and monitored in relative depth, so called 'emerging' mycotoxins have been identified. Similar to traditional mycotoxins, emerging mycotoxins are directly produced by fungi. Although some have been known for several decades already, emerging mycotoxins are still, however, less investigated and understood. One example of emerging mycotoxins is moniliformin (MON), which is also produced by some *Fusarium* species [8].

In addition to the free mycotoxins produced by toxigenic fungi (free parent compounds), mycotoxins can be modified in their chemical structure by biological or chemical processes [9–11]. Further, mycotoxins can be connected to the matrix, either by being physically entrapped or by covalent binding to matrix molecules. A proposal for a harmonized terminology of modified and matrix-associated mycotoxins was provided by Rychlik et al. [11]. Such forms can raise analytical challenges, which is particularly true for matrix-associated mycotoxins. For those, special treatments of the matrix, after extraction of free mycotoxins, are required to be able to extract the bound forms. However, free modified forms are, similar to emerging mycotoxins, often not covered by routine analysis.

Food processing, in general, is capable of affecting mycotoxins present in the raw materials. A reduction in mycotoxin concentrations might be caused by fractionation or (partial) degradation of the toxins, leading also to lower toxicity. However, often, lower mycotoxin levels (also) involve modification or binding mechanisms. In such cases, the resulting structures might still harbor unknown toxicity or might be (re)converted into a toxic form. Further, concentrations of free toxins can even increase during food processing by releasing mycotoxins from matrix components (if matrix-associated mycotoxins are present in the raw material and depending on the processing conditions). Tortillas are a traditional staple food for the Mesoamerican population and are increasing in popularity throughout the world [12], which also applies to related products, such as tostadas, tortilla chips, and maize chips. Their unique flavor is caused by an alkaline cooking of maize—a special processing procedure called nixtamalization. This process can cause several physicochemical changes in maize kernels and is capable of affecting mycotoxins. This review presents the current knowledge on mycotoxin changes during the process of tortilla production with a particular focus on the nixtamalization step. In doing so, changes in the concentration of free parent forms are considered, as well as their potential modification and the putative toxicological impacts.

2. Processes Involved in Nixtamalization and Tortilla Production

Nixtamalization describes an ancient food processing procedure developed and applied by indigenous Mesoamerican (e.g., Aztec and Mayan) civilizations [13], which is still used nowadays. It represents alkaline cooking of maize kernels. Traditionally, this is done using lime, which mainly consists of $Ca(OH)_2$. Classic nixtamalization also makes use of wood ash. In traditional nixtamalization (Figure 1), maize kernels are cooked in lime water followed by steeping at room temperature, which typically takes place overnight. During nixtamalization, the elevated pH and high temperature facilitate the softening of the endosperm and the release of the pericarp. After steeping, the cooking and steeping liquid, the so-called nejayote, is removed. The alkaline-cooked kernels (nixtamal) are then washed with water to remove excessive lime, as well as (part of) the loosened pericarp. The aleurone layer, i.e., the outermost layer of the endosperm that is rich in protein and vitamin B1–3, stays attached to the starchy endosperm. The aleurone layer also contributes to a reduction of protein and starch losses during cooking, steeping, and washing [14]. After washing, the nixtamal is stone-ground upon the addition of water to form a maize dough called masa. Small portions of masa are formed into balls that are flattened into thin discs. The so-formed tortillas are baked on a hot plate or in an oven. By frying, tortillas and tortilla strips can be further processed into tostadas and tortilla chips (or similar products). Additionally, masa can also be used to prepare maize chips. With respect to storage, masa can be dried and later remoistened for further processing. Moreover, a dry masa flour can be produced from low-moisture nixtamal by fine grinding under dry conditions. In this process, due to the low moisture, no release of starch granules from the protein matrix occurs compared with fresh masa production, in which nixtamal is ground at a high moisture level. Dry masa flour is often used in commercial tortilla production.

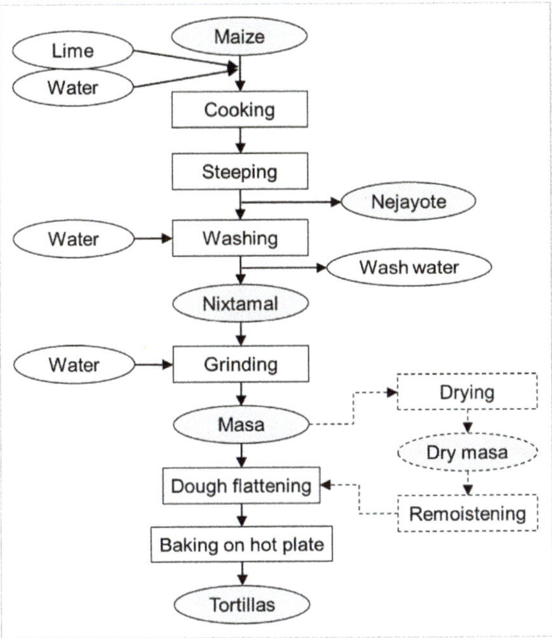

Figure 1. Scheme of typical steps and (by)products in traditional nixtamalization and tortilla production. The industrial production of tortillas often makes use of dry masa flour, which is made from dried nixtamal by fine grinding (not shown).

Alkaline cooking of maize causes several physical, as well as (bio)chemical, changes [14,15]. Some of those are associated with the enhanced nutritional value of the grain and are of particular importance in diets mainly relying on maize. The improved bioaccessibility of calcium and niacin (vitamin B3) are likely the most important of these changes. Thanks to the latter, pellagra—a niacin-deficiency disease typically related to maize- and sorghum-based nutrition—is not common in tortilla-eating countries. Moreover, nixtamalization can improve protein digestibility and can positively affect the protein quality of maize by partially adjusting the leucine-to-isoleucine disproportion [16]. In tortillas, few amino acids were found to be reduced—including leucine, which can act as an antagonist of isoleucine. Nonetheless, maize protein is, in general, deficient in lysine and tryptophan but relatively rich in the sulfur-containing amino acids methionine and cystine. Nixtamalization furthermore enhances the content of resistant starch, which is accompanied by a lower glycemic index [17,18]—a factor that is nowadays of special interest. Furthermore, traditional tortilla production is capable of lowering mycotoxin contaminations, as outlined below, which is of particular significance for a healthy cereal-based diet. Lime-cooked maize and products thereof are, moreover, characterized by a unique flavor, which contributes to the increasing popularity of such foods on a global level [12,19].

3. Aflatoxins during Nixtamalization and Tortilla Production

3.1. Impact on Aflatoxin Concentrations by Traditional Nixtamalization

Because aflatoxins, particularly AFB1, are a severe health threat and are often found in maize produced in warmer climates, strategies for reducing aflatoxin contaminations are of special importance. Interestingly, traditional nixtamalization is in several studies described to have a high potential for lowering aflatoxin concentrations (for details, see Table 1). The nejayote, the main waste byproduct, which typically also contains a solid fraction that mainly consists of maize tip cap, pericarp, and germ, can in return show a certain accumulation of aflatoxins.

The loss in aflatoxins and potential transfer to the lime water is also dependent on the type of aflatoxin [20,21]. In a study by Ulloa-Sosa et al. [21], AFB1 + AFB2 were less reduced (by around 40%–50% in nixtamal and masa) compared with AFG1 + AFG2 (by around 75%). The total aflatoxin concentration in nixtamal and masa was approximately reduced by 60%–65%. Remarkably, whereas most of the AFB1 + AFB2 were detected in the nejayote, no AFG1 or AFG2 could be detected in this matrix, pointing to a degradation or transformation into undetectable form(s) upon exposure to alkaline pH. Tortillas showed approximate reductions in total aflatoxins, AFB1 + AFB2, and AFG1 + AFG2 levels of 70%, 60%, and 80%, respectively. Additionally, de Arriola et al. [20] found AFG1 and AFG2 to be somewhat more reduced during nixtamalization than AFB1 and AFB2 (average reduction of AFB1, AFB2, AFG1, and AFG2 in masa approximately 93%, 90%, 98%, and 97%, respectively). Both studies were performed with fungal-inoculated kernels. Abbas et al. [22] analyzed the impact of the entire tortilla production process (covering nixtamalization using a 2% $Ca(OH)_2$ solution) on natural aflatoxin contaminations. Here, the AFB1 content was on average reduced by 40%, and the AFB2 content was reduced by only 28%. AFG1 and AFG2 were not detected in unprocessed maize (or in tortillas).

Enhancing the concentrations of $Ca(OH)_2$ for cooking and steeping would not necessarily cause a more pronounced aflatoxin reduction in tortillas [20,23]. Lime concentrations of >2% are even described to produce tortillas with organoleptic characteristics unsuitable for human consumption [20]. De Arriola et al. [20] found average reductions of total aflatoxin concentrations in masa and tortillas of 94% and 95%, respectively, at 0.6% lime without significant difference compared with the use of 1.87% lime. The experiments were done on highly aflatoxin-contaminated maize obtained by fungal inoculation. Moreover, the cooking procedure (i.e., cooking in an open kettle versus pressure cooking in an autoclave) made no significant difference regarding the change in total aflatoxin concentration [20]. When comparing the impact of five different nixtamalization processes on aflatoxin concentrations (see Table 1), Price et al. [23] found cooking, prolonged steeping, and washing of nixtamal to facilitate aflatoxin reductions in nixtamal, masa, and tortillas. Here, aflatoxin levels were reduced by approximately 50%–70% in tortillas compared with the naturally contaminated unprocessed maize (initial aflatoxin level of around 140 µg/kg). When using AFB1-spiked kernels (100 µg/kg) in previous experiments, the authors mentioned that the toxin was almost not detected after tortilla production [23].

Based on their experimental data, Moreno-Pedraza et al. [24] proposed a (traditional) process for nixtamalization and tortilla production, which is supposed to completely reduce AFB1 contaminations. The key steps are as follows: (i) cooking kernels in 1% lime at 90 °C for 45 min; (ii) further steeping in the alkaline solution at 25 °C for 18 h; (iii) removal of the nejayote followed by only one washing step to remove the pericarp but to preserve the alkaline conditions (approximately pH 10) of the nixtamal; (iv) resting the masa, which has been obtained by grinding the washed nixtamal and still features the high pH value at room temperature for 40 min; and (v) flattening and cooking of the tortillas. The authors found that the nixtamalization step, the resting of the alkaline masa for ≥30 min, and/or the tortilla baking completely or almost completely eliminated AFB1 levels (either low natural contamination or a 115–125 µg/kg spike in the masa). In this study, the alkaline treatment lead to the formation of at least two unidentified degradation or transformation products: one of 301.25 Da (molecular formula: $C_{17}H_{16}O_{15}$) and another of 325.33 Da (molecular formula: $C_{17}H_{18}O_5$) [24].

To follow the fate of aflatoxin during traditional nixtamalization, radio-labeled AFB1 was used [25]. Natural AFB1 contamination (at low and high levels) was found to be lowered by 97%–100% after traditional nixtamalization using a chromatographic method. When radio-labeled AFB1 was spiked in unprocessed maize, the loss in radioactivity amounted to 84% in the masa. The remaining radioactivity was detected in the nejayote and the washing liquids (with decreasing levels in later washings).

Whereas the aflatoxin retained in the masa appeared intact (detectable by an antibody-based method), aflatoxin in the liquid waste fractions was undetectable by means of chromatography [25].

3.2. Aflatoxin Reductions by Alternative Nixtamalization Processes

The abovementioned studies point to a high potential of traditional nixtamalization to lower aflatoxin concentrations. However, traditional nixtamalization is a time-consuming process characterized by a relatively high input of water and energy. Moreover, the nejayote is considered to be a highly problematic byproduct due to its excessive pH, its high content of organic and insoluble matter, and other factors [26], and it is usually disposed in landfills, not utilized [27]. Thus, to reduce water and energy inputs and the amount of wastewater, alternative processing procedures were considered. Such methods were not only tested regarding organoleptic and nutritional characteristics, but also in view of aflatoxin reductions as outlined below.

One promising alternative method for masa and tortilla production might be the extrusion of maize meal upon the addition of lime. At a lime concentration of 0.2%, based on maize meal, organoleptic properties were found to be comparable to traditionally produced masa/tortillas (using a 0.33% lime solution, which represented 1% lime based on maize meal). Total protein and lysine contents were similar (or only slightly lowered), but the tryptophan loss was much lower in extruded masa compared with traditional masa [28]. A higher nutritional value, accompanied by an elevated weight gain and protein efficiency ratio in a rat feeding trial, was, in addition, shown for tortillas produced with maize meal extruded at 0.15%–0.25% lime (relative to maize meal mass) compared with traditional tortillas produced with masa prepared with 2% lime based on kernel mass (= 0.67% lime solution) [29]. Elias-Orozco et al. [30] evaluated the alkaline extrusion process regarding aflatoxin reductions in naturally contaminated maize. Astonishing, in the highly contaminated batch, they found a high level of aflatoxin M1 (AFM1) in the raw maize, which was only around 20% less than the AFB1 level. It had been previously described that *Aspergillus* spp. are capable of producing AFM1 (and aflatoxin M2); however, usually relatively low amounts of AFM1, compared with AFB1 or total aflatoxins, are found in maize or in culture media [31–34]. Furthermore, they detected AFB1-dihydrodiol in the raw maize, which is formed via enzymatic oxidation of AFB1 followed by non-enzymatic hydrolysis. Tortillas produced after extrusion of maize meal lacking any lime showed reductions in AFB1, AFB1-dihydrodiol, and AFM1 levels of approximately 46%, 54%, and 20%, respectively. When lime was added at 0.3% relative to maize meal mass, reductions increased to approximately 74%, 70%, and 52%, respectively. At 0.5% lime, they amounted to 83%–89%. Moreover, the authors tested extrusion upon the addition of 0.75%–3% H_2O_2, alone or in combination with 0.3% lime. Similar to lime, adding H_2O_2 to the extrusion process can increase aflatoxin reductions in tortillas. However, the combination of lime and H_2O_2 showed no or minor benefits compared with lime alone, except regarding AFM1 reduction. Here, reduction was enhanced from approximately 52% (0.3% lime, only) to 61%–73% (0.3% lime + 0.75%–3% H_2O_2 based on maize meal). To compare, traditional nixtamalization of the highly contaminated maize resulted in AFB1, AFB1-dihydrodiol, and AFM1 reductions of 92%–94%. Similar results on the effect of extrusion treatments and traditional nixtamalization were found for a maize batch contaminated with AFB1 at a lower level; other aflatoxins were here not detected [30] (see Table 1 for details).

Pérez-Flores et al. [26] tested the use of a microwave for the nixtamalization process. For that, maize grits (obtained from fungal-inoculated kernels) were cooked in a minimized amount of lime water (0.5% $Ca(OH)_2$) in a microwave (for details, see Table 1). After steeping (3 h), no water removal or washing, which could have caused fractionation of mycotoxins, was indicated by the authors. The so-produced masa had 36%–82% lower aflatoxin concentrations than the maize kernels, with higher reduction at higher initial contamination levels. In tortillas, aflatoxin levels were lowered by 68%–84% of the initial amount. Tortillas produced by such microwave nixtamalization showed comparable physicochemical (moisture, pH, color) and technological properties (puffing, rollability, weight loss) as described for traditionally produced tortillas.

Minimization of water and energy input was also tested with a so-called ecological nixtamalization process [35]. Here, maize meal was mixed with a minimum amount of hot (92 °C) 0.375% lime solution for only 10 min. After steeping (2 h), the nixtamal was ground into masa without any water removal or washing steps. With such a process, AFB1 + AFB2 levels were lowered by 25%–40%, 13%–25%, and 61%–78% in the nixtamal, masa, and tortillas, respectively. Higher percentage reductions were, however, detected at lower initial contamination levels. Although applied to milled maize, the ecological nixtamalization was overall less effective in reducing aflatoxins compared with a traditional nixtamalization process applied to kernels of the same (fungal-inoculated) batch. The tested traditional nixtamalization covered a higher concentration of the lime water (1% lime), longer incubation times (70 min cooking, 12 h steeping), removal of the nejayote, and washing of the nixtamal (which typically also removes the loosened pericarp). Here, aflatoxin reductions amounted to 83%–92%, 87%–89%, and 90%–92% in nixtamal, masa, and tortillas [35].

Torres et al. [36] compared a traditional process (that included the cooking of kernels in lime water) with a commercial one. In the latter, whole maize kernels were mixed with lime and boiling water without further cooking (similar as described to the aforementioned ecological nixtamalization of maize meal [35]). In both processes tested by Torres et al. [36], the nejayote was removed after a 14 h steeping, and the nixtamal was washed twice. In the commercial process, the pH levels of the nixtamal, masa, and tortillas were lower compared with those in the traditional one (5.8–5.9 compared to 6.7–6.8). Moisture content was also slightly lower. Further, the loss of solids was reduced (4.1% in the commercial versus 6.8% in the traditional process). These factors likely contributed to the lower efficiency of aflatoxin reduction: The commercial processing reduced the level of total aflatoxins in tortillas by 30%, whereas the traditional tortilla production was more efficient (52% aflatoxin loss). Maize chips and tortilla chips showed aflatoxin reductions of 71% when using traditionally produced masa and 79%–85% upon use of masa produced with the tested commercial process. However, maize at different initial aflatoxin concentrations was applied to the two processes, which might also have affected the aflatoxin reduction efficiency. In this study, samples were acidified upon extraction (before filtration of suspended samples) to cause a reconversion of potential transiently transformed aflatoxins [36] (see below).

3.3. Potential Reconversion of Modified Aflatoxins

In general, besides the leaching of aflatoxins into liquid fractions, alkaline conditions can cause the opening of the lactone ring of aflatoxins (including AFB1), resulting in a loss of fluorescence of the molecules and thus a loss of fluorescence-based detection. Further, a strongly reduced toxicity and mutagenicity after cleavage of the lactone ring was described [37]. In nixtamalized maize (products), the lowered aflatoxin concentrations were found to be accompanied by lower mutagenicity and oxidative stress in vitro [23,38]. Vázquez-Durán et al. [38] showed that for extracts of raw maize, a more pronounced lipid peroxidation in kidney Vero cells occurred than for extracts of tortillas, which were produced from the raw maize by a microwave nixtamalization process (as described by Pérez-Flores et al. [26]) and had a 84% lower aflatoxin level. Further, no mutagenic toxicity was detected in the tortillas in the Ames test, but it was present in the unprocessed maize. Similarly, Price and Jorgensen [23] observed a reduced mutagenic potential for masa and tortilla samples compared with raw maize when testing different nixtamalization processes (although the number of revertants in the Ames test did not always correlate with the detected aflatoxin level regarding the extent of reductions).

The modification of aflatoxins during alkaline treatment is not necessarily permanent, however, and might be reversed upon exposure to acidic conditions, as present in monogastric digestive systems. Price and Jorgensen addressed this issue by acidifying the suspended samples in the course of aflatoxin extraction (original pH around 11; acidified: 5–6), mimicking acidification in the human stomach. In fact, in doing so, part of the undetectable modified aflatoxin(s) was reconverted into fluorescent form(s). After acidification, the total aflatoxin reduction in the tortillas amounted to only 20%–46%, instead of 48%–73% when lacking such a step. For masa, the reduction in fluorescent aflatoxin(s) was approximately 14%–56% without and 4%–29% with acidification. Additionally, Méndez-Albores et al. [39] showed that the reduction in aflatoxin concentrations by nixtamalization is partly reversible. After acidification, aflatoxins became, to some extent (approximately 5% of the concentration in raw maize), detectable in the dried nejayote, which originally had a pH of 12. When the extracts of the samples were acidified (initial pH of samples: 8.2–8.3), aflatoxin concentrations were around 57% and 34% higher compared with those in the alkaline extracts of masa and tortilla, respectively. However, compared with the raw maize, the aflatoxin levels were still very low with reductions of 78% in masa and 91% in tortillas (for the alkaline extracts, reductions amounted to 86% and 93%, respectively). Pérez-Flores et al. [26] found that only very low amounts of aflatoxins in the extracts of masa and tortillas were recovered by an acidification step. Here, the pH of masa and tortillas was again around 8.2 prior acidification (and adjusted to 3). Different from Price and Jorgensen [23], in the two latter studies [26,39], not the suspended samples, but the sample extracts were acidified. However, when treated with weak bases and during ammoniation, AFB1 was found to interact with matrix macromolecules, including non-protein fractions [40,41]. Hence, it is tempting to speculate that the extraction efficiency of the modified aflatoxin(s) is dependent on the pH of the matrix and that it is higher under acidic conditions. If this is true, matrix-associated aflatoxins potentially present in tortillas could be also released in the stomach.

In the study by Price and Jorgensen [23], acidification of the samples was found to be further capable of restoring mutagenicity in the Ames test. The mutagenic effects for the tortilla samples were even somewhat higher than for raw maize, which, however, contradicted the reduced aflatoxin concentrations that were observed in the acidified tortilla samples [23]. This might indicate the formation of additional mutagenic form(s) during tortilla production, which could also explain the rather low correlation between mutagenicity and aflatoxin concentration determined by the authors. In general, the efficiency and persistency of aflatoxin transformation/detoxification by elevated pH is dependent on several factors. Positive effects of temperature, time, and kernel moisture on AFB1 reduction under alkaline conditions were shown for ammoniation at atmospheric pressure [42] and under elevated pressure [43]. Differences in aflatoxin reductions depending on the processing procedures were also found for nixtamalization and tortilla production (Table 1). Further, initial contamination levels and type of contamination (contamination in the field, post-harvest contamination, spiking with pure standard) might affect mycotoxin reduction efficiencies. Such factors should be considered when assessing the aflatoxin loss in view of toxicity in alkaline-processed maize and products thereof, such as tortillas.

Table 1. Effect of alkaline cooking (nixtamalization) of maize kernels and of entire tortilla (chips) production on aflatoxin contents.

| Study No. in Figure 2 | Nixtamalization | | | | | Tortilla Baking on Hot Plate | Afla-toxin(s) | Initial Level in Raw Maize (μg/kg) [a] | Corrected Change (%) [b] | Comment(s) | Reference |
	Alkaline Solution	Additions on Orig. Maize Mass Basis	Cooking	Steeping	Washing of Nixtamal						
1	ca. 7.5% lime solution	167% water, ca. 12.5% lime	60 min	Only cool down after cooking	Not indicated	Performed (further details not provided)	Total AFs	Not provided (inoculated)	Nixtamal: ca. −65 [c] Nejayote: accum. Masa: ~−60 [c] Tortillas: ~−74 [c]	Overall: Relative quantification only and very approximate data	[21]
							AFB1 + AFB2	Not provided (inoculated)	Nixtamal: ~−51 [c] Nejayote: accum. Masa: ~−38 [c] Tortillas: ~−61 [c]	Most of the lost AFB was found in the nejayote	
							AFG1 + AFG2	Not provided (inoculated)	Nixtamal: ~−75 [c] Nejayote: n.d. Masa: ~−74 [c] Tortillas: ~−83 [c]	AFG1 + AFG2 were largely decomposed	
2	2% Ca(OH)₂ solution	Not specified	5 min (incl. stirring)	12 h	Thorough rinse with distilled water	110–120 °C, 7–8 min on each side	AFB1	417–476	Tortillas: −28 to −49 (mean: −40)	-	[22]
							AFB2	54–59	Tortillas: −17 to −40 (mean: −28)	-	
3	0.6%–1.87% lime solution	1480% water, 1–3% lime	95 °C, 40 min or 121 °C, 0.34 bar, 30 min	O/N	Several washes with tap water	180–250 °C (internal temp.: 94 °C), 1.5 min	Total AFs	5245–60,478 (inoculated)	Masa: ~−85 to −98 Tortillas: ~−92 to −99	No significant differences between different treatments	[20]
							AFB1	3265–37,696 (inoculated)	Masa: ~−89 to −96 Tortillas: ~−94 to −98		
							AFB2	512–9513 (inoculated)	Masa: ~−83 to −97 Tortillas: ~−88 to −98		
							AFG1	1236–11,131 (inoculated)	Masa: ~−97 to −99 Tortillas: ~−97 to −100		
							AFG2	226–1639 (inoculated)	Masa: ~−93 to −100 Tortillas: ~−96 to −100		

Table 1. *Cont.*

Study No. in Figure 2	Nixtamalization					Tortilla Baking on Hot Plate	Afla-toxin(s)	Initial Level in Raw Maize (μg/kg) [a]	Corrected Change (%) [b]	Comment(s)	Reference
	Alkaline Solution	Additions on Orig. Maize Mass Basis	Cooking	Steeping	Washing of Nixtamal						
4	0.33% lime solution	300% water, 1% lime	90 °C, 45 min	24 °C, 18 h	1 rinse with tap water	-	AFB1	0.68	Masa: ca. −100 (<LOD) [c]	-	[24]
						150 °C, 5 min on each side	AFB1	125 (spiked to masa)	Tortillas: ca. −90 to −100 [c]	Range is due to different detection methods	
5	Lime solution	1% lime (water: not specified)	94 °C, 50 min	17 h	2–3 washes	-	AFB1	37|251	Masa: −100|−97 [d]	-	[25]
							³H-AFB1	~200 (spiked)	Masa: −81 to −84 [d] Nejayote: accum. Washing water: accum.	Range is due to different detection methods	
6	0.33% Ca(OH)$_2$ solution	300% water, 1% Ca(OH)$_2$	20 min	15 h	Thorough rinse with water	1 min on each side	AF	135	Nixtamal: ~−34/~−20 * Masa: ~−56/~−29 * Tortillas: ~−54/~−31 *	-	[23]
7	As described above	As described above	As described above	As described above	As described above	As described above	AF	142	Nixtamal: ~−15/~−7 * Masa: ~−49/~−18 * Tortillas: ~−59/~−46 *	Steeping was performed before cooking!	
-	As described above	As described above	-	As described above	As described above	As described above	AF	145	Nixtamal: ~−17/~−6 * Masa: ~−14/~−4 * Tortillas: ~−54/~−42 *	-	
8	7.8% Ca(OH)$_2$ solution	160% water, 12.5% Ca(OH)$_2$	60 min	Left to cool down for 1 h	As described above	As described above	AF	142	Nixtamal: ~−15/~−4 * Masa: ~−19/~−4 * Tortillas: ~−48/~−20 *	-	
9	0.25% Ca(OH)$_2$ solution	300% water, 0.75% Ca(OH)$_2$	75 min	24 h	-	As described above	AF	142	Nixtamal: ~−36/~−7 * Masa: ~−46/~−4 * Tortillas: ~−73/~−23 *	-	

Table 1. *Cont.*

Study No. in Figure 2	Nixtamalization					Tortilla Baking on Hot Plate	Afla-toxin(s)	Initial Level in Raw Maize (µg/kg) [a]	Corrected Change (%) [b]	Comment(s)	Reference
	Alkaline Solution	Additions on Orig. Maize Mass Basis	Cooking	Steeping	Washing of Nixtamal						
10	0.33% lime solution	300% water, 1% lime	98 °C, 40 min	14 h	2 washes with 300% water (based on orig. maize mass)	Baking in a three-tiered, gas-fired oven for 39 s. Average temp. at the three levels: 177 °C, 233 °C, 453 °C	Total AFs	ca. 110	Nixtamal: ~−28 * Nejayote: n.d. * Masa: ~−44 * Tortillas: −52 * Maize chips: −79 * Tortilla chips: −85 *	Both processes: Corn chips and tortilla chips were prepared by frying masa and tortilla strips in oil at 190 °C for 2 and 3 min, respectively	[36]
11	As described above	As described above	No cooking, but manual mixing of maize and lime with boiling water	As described above	As described above	As described above	Total AFs	ca. 43	Nixtamal: ~−12 * Nejayote: accum. * Masa: ~−23 * Tortillas: −30 * Maize chips: −71 * Tortilla chips: −71 *		
12	1% lime solution	200% distilled water, 2% lime	85 °C, 70 min	22 °C, 12 h	1 wash with 200% tap water (based on orig. maize mass)	270 °C, in total 50–54 s on each side	AFB1 + AFB2	29|93 (inoculated)	Nixtamal: −92|−83 [c] Masa: −89|−87 [c] Tortillas: −92|−90 [c]	—	[35]
13	0.375% lime solution	80% distilled water, 0.3% lime	No cooking, but manual mixing of maize and hot lime water (92 °C) for 10 min	22 °C, 2 h; no removal of water indicated	-	As described above	AFB1 + AFB2	29|93 (inoculated)	Nixtamal: −40|−25 [c] Masa: −25|−13 [c] Tortillas: −78|−61 [c]	Use of maize meal (particle size: 800 µm)	

Table 1. *Cont.*

Study No. in Figure [2]	Nixtamalization					Tortilla Baking on Hot Plate	Afla-toxin(s)	Initial Level in Raw Maize (µg/kg) [a]	Corrected Change (%) [b]	Comment(s)	Reference
	Alkaline Solution	Additions on Orig. Maize Mass Basis	Cooking	Steeping	Washing of Nixtamal						
14	1% lime solution	300% distilled water, 3% lime	85 °C, 35 min	22 °C, 14 h	1 wash 200% with tap water (based on orig. maize mass)	270 °C, in total 50–54 s on each side	AFB1 + AFB2	678/680 ** (inoculated)	Nejayote: n.d./slight accum. ** Masa: ~−86/~−78 ** Tortillas: −93/~−91 **	Most aflatoxins appear to be degraded	[39]
15	0.5% Ca(OH)$_2$ solution	100% water, 0.5% Ca(OH)$_2$	Microwave (1650 W, 2450 Hz), 5.5 min	22 °C, 3 h; no water removal	-	270 °C, in total 54–55 s on each side	AFB1 + AFB2	22–141 (inoculated)	Masa: −36 to −82/~−34 to −81 ** Tortillas: −68 to −84/~−67 to −83 **	Use of maize grits. Higher reduction at higher initial concentration	[26]
16	1% lime solution	300% water, 3% lime	90–96 °C, 30 min	O/N	Several rinses with tap water	290 °C, in total 40–80 s on each side	AFB1	495|29	Tortillas: −94|~−92	-	
							AFB1-diol	30|-	Tortillas: ~−93|-		
							AFM1	402|-	Tortillas: −92|-		
-	Water, only	75% water	Extruder (low shear, single-screw, 35 rpm screw speed, 87 °C barrel temp.)	-	-	As described above	AFB1	495|29	Tortillas: ~−46|~−68	All extrusion treatments: Use of maize meal (particle size: 800 µm)	[30]
							AFB1-diol	30|-	Tortillas: −54|-		
							AFM1	402|-	Tortillas: −20|-		
17	0.4%–0.67% lime solution	75% water, 0.3%–0.5% lime	Extruder (for details see above)	-	-	As described above	AFB1	495|29	Tortillas: −74 to −85|~−100	Higher reduction at higher lime concentration	
							AFB1-diol	30|-	Tortillas: ~−70 to −89|-		
							AFM1	402|-	Tortillas: ~−52 to −83|-		

Table 1. *Cont.*

| Study No. in Figure 2 | Nixtamalization | | | | | Tortilla Baking on Hot Plate | Afla-toxin(s) | Initial Level in Raw Maize (μg/kg) [a] | Corrected Change (%) [b] | Comment(s) | Reference |
	Alkaline Solution	Additions on Orig. Maize Mass Basis	Cooking	Steeping	Washing of Nixtamal						
-	0.4% lime, 1%–4% H_2O_2 solution	75% water, 0.3% lime, 0.75%–3% H_2O_2	Extruder (for details see above)	-	-	As described above	AFB1	495|29	Tortillas: ~−67 to −78|~−100	Higher reduction at higher H_2O_2 concentration. At 0.3% lime + 3% H_2O_2: affection of taste	
							AFB1-diol	30|-	Tortillas: ~−68 to −84|-		
							AFM1	402|-	Tortillas: ~−69 to −81|-		

[a]: If not mentioned otherwise, maize was naturally contaminated. [b]: Unless indicated otherwise, the change in the mycotoxin concentration is corrected for change in moisture content. Negative values: reduction; positive values: increase. [c]: Here, it is not clear if the change in the mycotoxin concentration is corrected for change in moisture. [d]: Here, change in the mycotoxin concentration is supposed to be corrected for change in moisture. *: Here, aflatoxin detection involved acidification of the suspended sample. Values before the forward slash (if present) are derived from analyzing samples without acidification. **: Here, aflatoxin detection involved acidification of the extracts. Values before the forward slash are derived from analyzing extracts before acidification. ~: Approximate values that were calculated for this overview by using the data provided in the cited literature. |: Here, individual data of two batches are given and separated by this symbol. AFB1: aflatoxin B1; AFB1-diol: aflatoxin B1 dihydrodiol; AF: aflatoxin(s) not further specified in the cited study but likely total aflatoxins B1 + B2 + G1 + G2; total AFs: aflatoxins B1 + B2 + G1 + G2. accum.: accumulation; LOD: limit of detection; n.d.: not detected; O/N: overnight; orig.: original.

4. Fumonisins during Nixtamalization and Tortilla Production

4.1. General Impact on Fumonisin Concentrations and Fumonisin Hydrolyzation

Fumonisins are very water-soluble mycotoxins, which can thus leach into the liquid fraction during cooking and steeping procedures. Furthermore, an alkaline treatment can result in a hydrolysis of the *O*-acyl bonds of fumonisins, leading to the formation of hydrolyzed fumonisins. Sydenham et al. [44] found, upon steeping of maize kernels and maize meal in 0.1 M Ca(OH)$_2$ (at room temperature, under continuous stirring), a reduction in FB1 concentrations and an accumulation of fully hydrolyzed FB1 (HFB1; also referred to as aminopentol). For maize meal naturally contaminated with FB1, almost all of the mycotoxin was lost. Here, around 78%–89% of the FB1 was converted into HFB1, with 68%–72% being transferred into the steeping liquid and 11%–17% remaining in the alkali-treated maize meal. The latter contained only up to 9% of the FB1 level of untreated maize meal. In total, around 11%–25% of the FB1 was retained as FB1 or HFB1 in the maize meal. When treating whole kernels (also naturally contaminated) in the same manner, the reduction in FB1 concentration amounted to 76%–99%. After treatment, kernels were manually sorted by the extent of pericarp loss. Kernels with fully removed pericarp showed almost no FB1 left, and only approximately 4% of the FB1 was detected to be present as HFB1. In kernels with partly removed pericarp, approximately 7% of the initial FB1 was present as HFB1, and 24% remained in the parent form [44]. Accordingly, the removal of the nejayote and of maize pericarp would contribute to fumonisin reduction in nixtamalized maize.

Pilot-scale processing of naturally contaminated maize simulating commercial nixtamalization and tortilla (chips) production showed significant reductions in concentrations of FB1 and FB2 [45,46]. Although having a similar pattern, the extent of fumonisin reduction varied in both studies among individual runs, independent of initial concentration. Voss et al. [45] used different maize batches for five runs. Dombrink-Kurtzman et al. [46] examined the same maize batch in two runs but found nonetheless strong variations, particularly for FB2 (for details, see Table 2). FB1 reduction in nixtamal was in both studies accompanied by an accumulation of HFB1 in the steeping and washing liquids. Voss et al. [45] detected, besides HFB1, some partially hydrolyzed FB1 (PHFB1) in the raw maize. However, this compound did not accumulate in the nejayote but, if present, decreased over time. The decrease in FB1 and PHFB1 was accompanied by an increase in the fully hydrolyzed form. In general, hydrolyzation particularly takes place in the nejayote, which typically has a pH of ≥11. Palencia et al. [47] found the molar ratio of HFB1 to FB1 to be 21 in the nejayote but around 1 in wash water, masa, and tortillas. The overall transfer of fumonisins to the nejayote amounted in the study of Voss et al. [45] to approximately 45% of the total initial amount (on a molar basis) of FB1, PHFB1, and HFB1. Additional amounts were detected in the washing water. Dombrink-Kurtzman et al. [46] described the liquid fractions to contain on average of 76% of the initial FB1: 72.5% of FB1 was converted into HFB1 and 3.5% remained as FB1. The study also indicated the potential for further lowering of FB1 and FB2 levels during masa/tortilla production, in case they were somewhat less reduced in the nixtamal. FB1 and FB2 reductions in tortillas amounted in both runs to 88%–92% and 71%–91% compared with the levels in unprocessed maize, respectively. However, in one of the runs, the nixtamal showed reductions of around 75% for FB1 and only 20%–30% for FB2 [46] (Table 2).

The potential impact of tortilla baking on fumonisin reductions also became obvious in a study that tested a microwave nixtamalization process using maize grits. Here, total fumonisin levels were not significantly lowered in masa. This was in accordance with the tested processing procedure, because no removal of nejayote or washing of nixtamal was indicated by the authors. Different from masa production, the baking of tortilla lowered fumonisins by approximately one half. The reduction during the heat treatment was likely facilitated by the high pH of masa and tortillas (i.e., around 8.1–8.3). The tested physicochemical and technological characteristics were similar to those described for traditionally produced tortillas [48].

The fate of fumonisins was also investigated during commercial processing into tortilla (chips) [49,50]. Scudamore et al. [49] analyzed industrial tortilla chip production in United Kingdom

(UK) plants, involving mixing of a maize flour dough followed by sheeting, cutting, baking, and frying. Alkaline conditions were, however, not indicated. However, because nixtamalization contributes to the typical flavor of tortillas and tortilla chips, we assume that dry masa flour or a similar ingredient was involved in the commercial production process. When analyzing 11 runs (that comprised two different compositions of maize flour mixtures), FB1 + FB2 were lowered by 32%–78% on the product 'as is' basis (average: 59%). Because the moisture content of the chips is usually more or less comparable to that of dry maize ingredients, a similar fumonisin reduction would apply when related to dry weight. Commercial tortilla production in Texas was studied by De La Campa et al. [50]. Here, the reduction in FB1 levels was high overall and ranged from 80% to 100% in masa and from 83% to 100% in tortillas. Production conditions strongly differed between the four processing plants, regarding, for example, lime concentration and cooking time.

In experimental studies, De La Campa et al. [50] further investigated the impact of these factors at different initial FB1 levels using fungal-inoculated maize. In doing so, they found a positive impact of lime concentration (when testing lime solutions of around 0.25%–1.6%) on FB1 reduction. This effect was independent of the initial FB1 concentration, which also had a significant effect on FB1 half-life. Regarding boiling time (15 versus 60 min), the authors mentioned that this factor had no apparent effect, but data were not shown [50]. Additionally, De Girolamo et al. [51] described the low effect of cooking time on the hydrolyzation of fumonisins, when comparing cooking times of 15, 30, and 60 min. In this study, which tested lime solutions with concentrations of around 0.33% and 1.67%, nixtamalization lowered mean FB1 + FB2 levels in masa by 26%–48%. Interestingly, the same process lacking lime resulted in a somehow stronger FB1 + FB2 reduction. Here, PHFB (PHFB1 + PHFB2) levels in masa were also lowered, but the loss was not accompanied by the formation of HFBs (HFB1 + HFB2). Reductions are likely solely caused by the leaching of fumonisins into the liquid fractions. By contrast, the use of lime provoked the formation of (partially) hydrolyzed forms of FB1 and FB2. Hydrolyzation was again more pronounced at higher lime concentration [51].

4.2. Potential Further Transformations of Fumonisins

On the one hand, De Girolamo et al. [51] discovered that alkaline cooking can somehow facilitate the release of bound fumonisins. In their study, the total mass of FB1 + FB2, PHFBs, and HFBs recovered after nixtamalization (also including the liquid waste fractions) exceeded the initial mass by around 50%–80%. In the course of nixtamalization, (part of) the released matrix-associated fumonisins were suggested to be hydrolyzed. Although the water-cooked maize showed a higher reduction in (free) FB1 + FB2 than the alkali-cooked maize, matrix-associated fumonisins would still be present. Moreover, bound fumonisins in food (and feed) products can, in general, increase health concerns, because free toxins might be released during digestion. Promoting the release from the matrix followed by hydrolyzation of fumonisins could contribute to a detoxification by nixtamalization. Different from FB1 and FB2, no (liver) cancer-promoting activity or weight loss was found for HFB1 and HFB2 in rats [52]. In contrast, in vitro tests showed a higher toxicity on primary rat hepatocytes in this study. Hence, it was concluded that the hydrolyzed fumonisins are not adsorbed from the gut [52]. A lower or lacking hepatic, intestinal, and neural toxicity of the hydrolyzed form compared with the parent compound was also shown in pigs and mice [53,54], although an impact on sphingolipid metabolisms in vivo was demonstrated at a high dose of HFB1 [54]. Inhibition of ceramide synthase and disruption of sphingolipid metabolism is the critical biochemical effect underlying fumonisin cytotoxicity.

On the other hand, hydrolyzation might also favor interaction with other compounds, including matrix macromolecules. Interestingly, Park et al. [55] were able to detect matrix-associated fumonisins in some retail tortilla chip samples. However, when analyzing retail samples, it cannot be excluded that the forms were already present in the raw material. To address this question, Burns [56] investigated a nixtamalization process by applying the detection method developed by Park et al. [55]. In doing so, a significant increase in protein-bound and other matrix-associated FB1 during nixtamalization was demonstrated. When maize kernels were processed in the same manner, but lacking lime, no significant change was observed in the concentration of total matrix-associated FB1 [56]. A reduction in recoverable (H)FB1 was described for experimentally produced and extruded masa flour [57]. However, here, the underlying mechanisms (degradation, binding, or modification to undetected free forms) remained unknown.

N-(carboxymethyl)-FB1 was previously shown to be formed under alkaline conditions at elevated temperatures by using pure FB1 incubated overnight with D-glucose [58] and also when heating HFB1 with D-glucose [59]. Interaction with glucose during extrusion cooking of maize resulted in a strong reduction in fumonisin-induced toxicity in rats [60]. However, when analyzing nixtamalization and tortilla chip production mimicking commercial processing, no indications were given for a (relevant) formation and accumulation of fumonisin–sugar adducts, namely *N*-(carboxymethyl)-FB1 and *N*-(1-deoxy-D-fructos-1-yl)-FB1 [45]. By contrast, Park et al. [61] could detect *N*-fatty acyl fumonisins in a tortilla chip sample (in 1 out of 38 retail samples), indicating a potential formation of those modified fumonisins in alkali-treated and fried maize products. In vitro studies implicate a high toxicity of several *N*-fatty acyl fumonisins. In addition, such modified forms can be more rapidly taken up and accumulated in human/animal cells than FB1 (for an overview, see [4]). However, further studies are required to obtain more information on the toxicity of *N*-acetylated fumonisins and their occurrence in foods. The same is true for the interaction of fumonisins with other molecules that potentially takes place during nixtamalization and the possible contribution to fumonisin-related toxicity.

Using bioassays and feeding trials, several studies indicate a reduced toxicity of FB1-contaminated maize raw material after being processed by nixtamalization [47,62–64]. To analyze (potential) kidney damage, in addition to histological analysis, sphinganine can be used as a biomarker for fumonisin-induced ceramide synthase inhibition. Palencia et al. [47] detected reduced accumulation of sphinganine in cell lines treated with extracts of tortillas compared with those treated with extracts of raw maize. This was in conjunction with lowered FB1 levels. Here, the sum of FB1 and HFB1 in tortillas (on molar basis) was half of the initial FB1 level detected in the raw maize. The toxic potential was found to be lowered by 60% for extracts of tortillas compared with extracts of unprocessed maize [47]. Similarly, in feeding trials on rats, kidney sphinganine and sphingosine concentrations were not increased or less increased in rats fed a diet containing nixtamalized maize (meal) compared with those fed non-nixtamalized maize (meal) [62,63]. In both studies, nixtamalization was performed with a 1.2% lime solution. Voss et al. [63] used raw maize with three different FB1 contamination levels. Rats that ate a diet containing nixtamalized maize showed no or only week symptoms of nephropathy. This was much different from when the diet contained uncooked maize. Burns et al. [62] additionally included a mock-nixtamalization control (i.e., cooking of maize meal without lime). Similar to the findings of De Girolamo et al. [51], this procedure also lowered the FB1 level, but much less hydrolyzation took place compared with cooking with lime (for details, see Table 2). Both nixtamalization and mock-nixtamalization strongly reduced kidney damage, as well as renal toxicity (evaluated by number of apoptotic tubule cells), compared with uncooked maize meal [62]. Due to the clearly reduced toxic effects caused by nixtamalized maize (products), a significant formation of matrix-associated FB1 or unknown free fumonisin forms that contributed to toxicity was not indicated in these two studies. This differed from a former rat feeding trial performed by Hendrich et al. [64] using highly FB1-contaminated maize (obtained by fungal inoculation). Here, although nixtamalization was able to lower toxicity in some cases, a more pronounced effect would be expected in view of the high loss in FB1 (approximately 98%–100%). When considering the molecular weights, the formed HFB1 amounted to approximately 60%–72% of the initial FB1. In this study, it became further obvious that the nutritional status was capable of impacting toxicological effects caused by fumonisins present in the non-nixtamalized and nixtamalized maize. Hence, more research is needed regarding the potential formation and occurrence of so far undetected and/or unknown toxic fumonisin form(s) in alkali-cooked maize.

Table 2. Effect of alkaline cooking (nixtamalization) of maize kernels and of entire tortilla (chips) production on fumonisin contents.

Study No. in Figure 2	Nixtamalization					Tortilla Baking on Hot Plate	Fumo-nisin(s)	Initial level in Raw Maize (µg/kg) [a]	Corrected Change (%) [b]	Comment(s)	Reference
	Alkaline Solution	Additions on Orig. Maize Mass Basis	Cooking	Steeping	Washing of Nixtamal						
18	~0.37% lime solution	~1290% water, ~4.8% lime	100 °C, 5 min in a steam kettle (in a perforated nylon bag)	15 h	Wash with water	Baking in a gas-fired oven with three moving tiers (further details not provided)	FB1	8790	Nixtamal, unwashed: ~−76|~−97 Nixtamal, washed: ~−73|~−96 Nejayote: n.d.|slight accum. Washing water: n.d.|slight accum. Masa: ~−87|~−92 Tortillas: ~−88|~−92 Tortilla chips: n.a.|~−94	Overall: Data of two production runs are given; tortilla chips were produced in only one run by frying in oil at 190 °C for 60 s. Of the initial FB1, a total of 62%–90% was recovered as FB1 or HFB1 [e] from the nejayote, mostly as HFB1	[46]
							HFB1	Probably n.d.	(Slight) accum. in all intermediate products and (by)products, highest accum. in nejayote		
							FB1 + HFB1	~12,000 nmol/kg	Nixtamal, unwashed: ~−48|~−85 [f] Nixtamal, washed: ~−48|~−87 [f] Masa: ~−70|~−78 [f] Tortillas: ~−72|~−82 [f] Tortilla chips: n.a.|~−86 [f]		
							FB2	1,970	Nixtamal, unwashed: ~−32|~−97 Nixtamal, washed: ~−21|~−92 Nejayote: n.d.|n.d. Washing water: n.d.|n.d. Masa: ~−78|~−94 Tortillas: ~−71|~−91 Tortilla chips: n.a.|~−90	-	

Table 2. *Cont.*

Study No. in Figure 2	Nixtamalization					Tortilla Baking on Hot Plate	Fumo-nisin(s)	Initial level in Raw Maize (µg/kg) [a]	Corrected Change (%) [b]	Comment(s)	Reference
	Alkaline Solution	Additions on Orig. Maize Mass Basis	Cooking	Steeping	Washing of Nixtamal						
19	Processing in a pilot plant according to commercial procedures including alkaline cooking, steeping, and washing of maize kernels, as well as baking and deep frying of masa to produce tortilla chips (details not provided)						FB1	220–46,500	Nejayote: slight accum. Masa: −88 to −94 [d] Baked tortilla chips: −76 to −90 [d] Fried tortilla chips: −36 to −78 [d]	Ca. 34% and 45% of the initial FB1 [f] was detected as FB1, PHFB1, or HFB1 in the masa and nejayote, respectively. No indications for significant fumonisin–sugar adduct formation/accumulation	[45]
							PHFB1	n.d.–1340	Nejayote: n.d. or slight accum. Masa: ~−33 [d] to slight accum.		
							HFB1	n.d.–950	Nejayote: accum. Masa: no change to slight accum.		
							FB2	(Not provided)	Masa: ca. −89 to −94 [d] Baked tortilla chips: ca. −89 to −94 [d]	-	
-	-	-	-	-	-	Baking in a tortilla oven at 260 °C, 20 s	FB1 + FB2	100–281	Tortilla chips: 'as is': −32 to −78 (mean: −59)	Use of maize flour mixtures. Commercial processing. Tortilla chips were prepared by frying tortilla strips at 170–175 °C for 40 s	[49]
20	~1% lime solution	~77% water, ~0.8% lime	7 min	18 h	No information	250 °C	FB1	1001	Cooked maize: −56 [c] Nixtamal: −83 [c] Masa: −80 [c] Tortillas: −83 [c]	Commercial processing. Cooked maize was sampled before steeping	[50]
21	~0.7% lime solution	~200% water, ~1.5% lime	150 min	16 h	No information	250 °C	FB1	681	Cooked maize: −73 [c] Nixtamal: −80 [c] Masa: −85 [c] Tortillas: −89 [c]		
22	~0.56% lime solution	~160% water, ~0.9% lime	10 min	16 h	No information	375 °C	FB1	1441	Cooked maize: −83 [c] Nixtamal: −88 [c] Masa: −100 [c] Tortillas: −100 [c]		
23	~1.1% lime solution	~70% water, ~0.8% lime	120 min	18 h	No information	232 °C	FB1	1653	Cooked maize: −78 [c] Nixtamal: −88 [c] Masa: −100 [c] Tortillas: −100 [c]		

Table 2. *Cont.*

Study No. in Figure 2	Nixtamalization					Tortilla Baking on Hot Plate	Fumo-nisin(s)	Initial level in Raw Maize (µg/kg) [a]	Corrected Change (%) [b]	Comment(s)	Reference
	Alkaline Solution	Additions on Orig. Maize Mass Basis	Cooking	Steeping	Washing of Nixtamal						
24	~0.25% lime solution	200% water; ~0.5% lime	60 min	18 h	Rinse in 430% water (based on orig. maize mass)	Baking over an iron grill at 190–200 °C for ca. 4 min	FB1	150–11,800 (inoculated)	Masa: −46 to −99 [c] Tortillas: −26 to −98 [c]	Higher reduction at higher initial FB1 level	
-	~0.25–1.6% lime solution	200% water; ~0.5–3.2% lime	15 or 60 min	18 h	As described above	As described above	FB1	150–11,800 (inoculated)	Masa/Tortillas: up to −100 [c]	Higher reduction at higher lime concentration (and at higher initial FB1 level)	
25	0.33 or 1.67% lime solution	300% distilled water, 1% or 5% lime	90 °C, 15–60 min	17 h	2 rinses with 200% tap water (based on orig. maize mass)	-	FB1 + FB2	6,480–8,930	Masa: −26 to −48 [c]	Of the initial FB1 + FB2[f], a total of 64%–86% retained as parent form, PHFB, or HFB in the masa	[51]
							PHFB1 + PHFB2	110–260	Masa: no change to accum.	Increase at higher lime concentration	
							HFB1 + HFB2	n.d.	Masa: accum.	Higher accum. at higher lime concentration	
-	Distilled water, only	300% distilled water	As described above	As described above	As described above	-	FB1 + FB2	5230–20,380	Maize dough: −45 to −78 [c]	Here, a total of 21%–55% of the initial FB1 + FB2[f] retained as parent form, PHFB, or HFB in the maize dough	
							PHFB1 + PHFB2	140–510	Maize dough: −29 to −75 [c]		
							HFB1 + HFB2	n.d.	Maize dough: n.d.		
26	0.5% Ca(OH)₂ solution	100% water, 0.5% Ca(OH)₂	Microwave (1650 W, 2450 Hz), 3.75 min	25 °C, 3.5 h; no removal of water indicated	-	270 °C, in total 54–55 s on each side	FB1 + FB2 + FB3	2137 (inoculated)	Masa: −6 (n.s.) [c] Tortillas: −54 [c]	Use of maize grits	[48]
27	1.2% lime solution	Not specified	95–100 °C, 55 min	14 h	Wash with ca. 300% tap water	-	FB1	239,000 (inoculated)	Nixtamal: −83 [c] Nejayote: slight accum.	Pericarp removal during washing was avoided	[57]
							HFB1	n.d.	Nixtamal: accum. Nejayote: accum.		
							FB1 + HFB1[e]	239,000	Nixtamal: −39 [c] Nejayote: accum.		
28	~1.3% lime solution	325% water, ~4.1% lime	ca. 105 min	15 h	3 rinses with 275% water (based on orig. maize mass)	170–212 °C, ca. 3.5 min	FB1 + HFB1 [e]	ca. 38,100	Tortillas: −46 [f]	-	[47]

Table 2. *Cont.*

| Study No. in Figure 2 | Nixtamalization | | | | | Tortilla Baking on Hot Plate | Fumonisin(s) | Initial level in Raw Maize (µg/kg) [a] | Corrected Change (%) [b] | Comment(s) | Reference |
	Alkaline Solution	Additions on Orig. Maize Mass Basis	Cooking	Steeping	Washing of Nixtamal						
29	1.2% Ca(OH)$_2$ solution	750% water, 9% lime	90–100 °C, 60 min	O/N	3 rinses with 750% water (based on orig. maize mass)	-	FB1	~23,314	Nixtamal: ~−90 [c]	Of the initial FB1 [f], a total of ~47% retained as FB1, HFB1, or bound (H)FB1 in the nixtamal	[56]
							HFB1	~329	Nixtamal: ca. +1400 [c]		
							Protein-bound (H)FB1	~69 (recovered as HFB1)	Nixtamal: ~+476 [c]		
							Total bound (H)FB1	~89 (recovered as HFB1)	Nixtamal: ~+673 [c]		
-	Water, only	750% water	As described above	As described above	As described above	-	FB1	~23,314	'Mock-nixtamal': ~−53 (n.s.) [c]	Here, a total of ~44% of the initial FB1[f] retained as FB1, HFB1, or bound (H)FB1 in the 'mock-nixtamal'	
							HFB1	~329	'Mock-nixtamal': ~−60 (n.s.) [c]		
							Protein-bound (H)FB1	~69 (recovered as HFB1)	'Mock-nixtamal': no change [c]		
							Total bound (H)FB1	~89 (recovered as HFB1)	'Mock-nixtamal': ~+9 (n.s.) [c]		
30	1.2% Ca(OH)$_2$ solution	1,200% water, 14.4% lime	90–100 °C, 60 min	O/N	3 washes with 1200% distilled water (based on orig. maize mass)	-	FB1	9080 (inoculated)	Nixtamal: −77	Raw material: ground maize used as fungal growth medium. Nixtamal was prepared for a feeding trial. Concentrations and changes are given for the mixed diet	[62]
							HFB1	250	Nixtamal: ~+408		
							FB1 + HFB1	13,200 nmol/kg	Nixtamal: ~−54 [f]		
-	Water, only	1,200% water	As described above	As described above	As described above	-	FB1	9080 (inoculated)	'Mock-nixtamal': −87		
							HFB1	250	'Mock-nixtamal': ~+120		
							FB1 + HFB1	13,200 nmol/kg	'Mock-nixtamal': ~−77 [f]		
31	1.2% Ca(OH)$_2$ solution	300% water, 3.6% lime	80–100 °C, 60 min	O/N	One wash with ca. 300% water (based on orig. maize mass)	-	FB1	45,200–48,000 (inoculated)	Nixtamal: ~−98 to −100	Nixtamal was prepared for a feeding trial. Concentrations and changes are given for the mixed diet	[64]
							HFB1	n.d.	Nixtamal: accum.		
							FB1 + HFB1	~62,600–66,500 nmol/kg	Nixtamal: ~−58 to −70 [f]		

[a]: If not mentioned otherwise, maize was naturally contaminated. [b]: Unless indicated otherwise, the change in the mycotoxin concentration is corrected for change in moisture content. Changes on a product 'as is' (wet weight) basis are indicated. Negative values: reduction; positive values: increase. [c]: Here, change in the mycotoxin concentration is supposed to be corrected for change in moisture. [d]: Here, it is not clear if the change in the mycotoxin concentration is corrected for change in moisture. [e]: As equivalent to parent form. [f]: On molar basis. ~: Approximate values that were calculated for this overview by using the data provided in the cited literature. |: Here, individual data of two production runs are given and separated by this symbol. FB1/2: fumonisin B1/2; HFB1/2: hydrolyzed fumonisin B1/2; PHFB1/2: partially hydrolyzed FB1/2. accum.: accumulation; n.a.: not analyzed; n.d.: not detected; n.s.: not significant; O/N: overnight; orig.: original.

5. Other Mycotoxins during Nixtamalization and Tortilla Production

Little data were found on mycotoxins other than aflatoxins and fumonisins during nixtamalization and tortilla production (Table 3). Abbas et al. [65] experimentally produced tortillas (including traditional nixtamalization with 2% Ca(OH)$_2$) using two batches of maize naturally contaminated with the *Fusarium* toxins ZEN and DON, as well as maize spiked with the purified toxins (by injection into the embryos). Here, no difference in the percentage reduction of mycotoxin was obvious depending on the type of contamination. For ZEN, the initial levels that were present as *trans*-ZEN were lowered by 59%–100%. For the two maize samples with the highest concentrations (one spiked and one naturally contaminated one), some ZEN (<0.4% of the total amount) could be detected in the nejayote. Further, some isomerization from *trans*-ZEN to *cis*-ZEN took place for these maize samples. However, most of the ZEN was degraded into undetectable form(s), and it was supposed that the alkaline treatment attacked the lactone ring of ZEN. Whether this transformation would be stable under acidic conditions was not addressed [65]. For DON, reductions amounted to 72%–82%. The naturally contaminated maize batches contained, in addition, the acetylated form 15-acetyl-DON, which was completely destroyed in tortillas. Neither DON nor 15-acetyl-DON could be detected in the nejayote.

The potential to lower ZEN and DON by alkaline steeping of maize was also shown when using 0.1 M sodium carbonate. Here, steeping of raw maize kernels at 22 °C for 24 h lowered ZEN and DON by around 45% and 70%, respectively. An extended steeping over 72 h reduced the concentrations by 88% and 95%, respectively [66]. The baking and frying steps in commercial tortilla chip production were analyzed by Scudamore et al. [49] regarding mycotoxin changes on a product 'as is' basis. In the (probably alkaline) maize flour mixture(s) used to prepare the dough for tortillas, ZEN was present at low levels only, and the change during processing was very variable. However, if the initial ZEN level was higher than 13 µg/kg, the reduction amounted to 35%–64%. If the initial level was below 9 µg/kg, the detected change ranged from a 7% reduction to a 116% increase. This was probably caused by difficulties in representative sampling of industrial processes. DON levels in the tortilla chips were on average lowered by 32%, with the highest reductions at the highest initial levels. The sensitivity of DON towards food production processes that involve alkaline additives was also observed in the production of bakery wares and during the cooking of noodles (for an overview, see [67]).

The reduction of the emerging mycotoxin MON during tortilla production was studied by Pineda-Valdes et al. [68]. In pilot-scale experiments, MON was reduced by 97% after cooking of maize kernels in a 0.25% lime solution. After steeping or further processing, MON could not be detected anymore. When determining laboratory-scale processing using fungal-inoculated maize with an around 10-fold higher initial concentration, MON was lowered by 54% during cooking. After steeping and washing, MON reduction accounted for 64% and 69%, respectively. In masa and tortillas, the loss was around 70%. MON was not detected in any of the liquid fractions [68], although it is characterized by low molecular size and high water solubility [8]. Thus, MON might have been either modified into undetected form(s) during the 20 min alkaline cooking step or was degraded due to the action of high temperature and/or high pH. In a former study, Pineda-Valdes and Bullerman [69] demonstrated an affection of MON at elevated pH and temperature. Heating to 100 °C in an aqueous environment with pH 10 for 60 min lowered MON by around one half. However, after 20 min of cooking, the MON loss amounted to less than 20%. Therefore, nixtamalization of MON-contaminated maize showed a relatively high efficiency in reducing the concentration of this emerging mycotoxin, probably by a pH of >10 of the lime water.

Table 3. Effect of alkaline cooking (nixtamalization) of maize kernels and of entire tortilla production on contents of mycotoxins other than aflatoxins and fumonisins.

Study No. in Figure 2	Nixtamalization					Tortilla Baking on Hot Plate	Myco-toxin(s)	Initial Level in Raw Maize (µg/kg) [a]	Corrected Change (%) [b]	Comment(s)	Reference
	Alkaline Solution	Additions on Orig. Maize Mass Basis	Cooking	Steeping	Washing of Nixtamal						
32	2% Ca(OH)₂ solution	Not specified	5 min (incl. stirring)	12 h	Thorough rinse with distilled water	110–120 °C, 7–8 min on each side	ZEN	230|4,230	Nejayote: n.d.|weak accum. Tortillas: −100|−59	-	[65]
							DON	3,280|12,260	Nejayote: n.d.|n.d. Tortillas: −82|−72		
							15-acetyl-DON	1,490|9,830	Nejayote: n.d.|n.d. Tortillas: −100|−100		
							ZEN	750|3,620 (spiked)	Nejayote: n.d.|weak accum. Tortillas: −71|−74	Mycotoxins were injected into the maize embryos	
							DON	850|4460|8250 (spiked)	Nejayote: n.d.|n.d.|n.d. Tortillas: −82|−72|−74		
-	-	-	-	-	-	Baking in a tortilla oven at 260 °C, 20 s	ZEN	4.5–8.7|19–24	Tortilla chips: 'as is': +116 to −7 (mean: +32)|−35 to −64 (mean: −49)	Use of maize flour mixtures. Commercial processing. Tortilla chips were prepared by frying tortilla	[49]
							DON	47–466	Tortilla chips: 'as is': +28 to −76 (mean: −32)		
33	0.25% lime solution	400% water, 1% lime	88 °C, 20 min	16 h	Two rinses with 250% water (based on orig. maize mass)	Baking in a gas-fired oven at ca. 365 °C for ca. 3 min	MON	1420	Cooked maize: −97 [c] Nixtamal: −100 [c] Nejayote: n.d. Wash water: n.d. Masa: −100 [c] Tortillas: −100 [c]	Pilot-scale process	[68]
34	0.25% lime solution	400% water, 1% lime	88 °C, 20 min	16 h	Two rinses with 125% water (based on orig. maize mass)	ca. 250 °C, 3 min on each side	MON	17,640 (inoculated)	Cooked maize: −54 [c] Nixtamal (before washing): −64 [c] Nixtamal (washed): ~−69 [c] Nejayote: n.d. Wash water: n.d. Masa: ~−71 [c] Tortillas: ~−70 [c]	Laboratory-scale	

[a]: If not mentioned otherwise, maize was naturally contaminated. [b]: Unless indicated otherwise, the change in the mycotoxin concentration is corrected for change in moisture content. Changes on a product 'as is' (wet weight) basis are indicated. Negative values: reduction; positive values: increase. [c]: Here, it is not clear if the change in the mycotoxin concentration is corrected for change in moisture. |: Here, data of different batches are separated by this symbol. ~: Approximate values that were calculated for this overview by using the data provided in the cited literature. DON: deoxynivalenol, MON: moniliformin; ZEN: zearalenone. accum.: accumulation; n.d.: not detected; orig.: original.

6. Conclusions and Outlook

Nixtamalization and tortilla baking can affect mycotoxins in different ways, including physical and chemical action: (1) Water-soluble mycotoxins can leach into the liquid fractions during cooking, steeping, and washing. (2) Mycotoxins present in the pericarp, tip cap, and germ are removed when these tissues are (partly) separated by thorough washing of the nixtamal. (3) Action of high pH and elevated temperature during cooking and baking can result in degradation, modification, and/or binding or release of mycotoxins.

For traditional nixtamalization, a high potential to lower free parent forms of mycotoxins is described (see also Figure 2). Aflatoxin concentrations of raw maize were found to be lowered by around 15%–85% and 20%–100% in the nixtamal and masa, respectively. Tortillas mostly showed aflatoxin reductions of 50%–100%. For FB1, the reduction mainly amounted to around 75%–100% in nixtamal, masa, and tortillas. For ZEN, DON, and MON, reductions of around 60%–100%, 70%–80%, and 70%–100% are described. However, only very limited data is available regarding maize mycotoxins other than aflatoxins and fumonisins. More data on such toxins would help to evaluate the benefits of alkali-processed maize in more detail.

Besides reduction in the free parent forms, modification of mycotoxins can occur, and interaction with matrix compounds can be altered. To analyze such processes during nixtamalization, intense efforts have already been undertaken to establish and optimize appropriate detection methods. Although it must be noted that, when analyzing food matrices, which can harbor several challenges, analytical recovery must in general be taken into account, and data should be corrected accordingly (which was often not done or at least not mentioned for the data presented in the current review). Careful conclusions on the reduction factors of mycotoxins must certainly also take a reasonable contribution of variability and uncertainty into account. Furthermore, the stability of the present forms and their bioavailability need to be considered when analyzing toxicological impacts. In addition, precise knowledge on the critical parameters in nixtamalization and tortilla production is important to optimize production procedures to furthermore reduce potential health risks to the consumers, e.g., by reconversion of aflatoxins after consumption.

In general, further research is needed to evaluate possible modifications and matrix–mycotoxin interactions during nixtamalization, as well as the occurrence and potential toxicity of the formed structures in the final food items. In doing so, a possible reconversion and/or release of parent forms in the gastrointestinal tract, as well as by activity of the gut microflora, need to be considered. Reliable analytical data would be the basis for precise understanding of the processes and the factors in mycotoxin reduction. Moreover, knowledge on the fate of mycotoxins and their toxicity is required to evaluate possible utilization strategies for the nejayote.

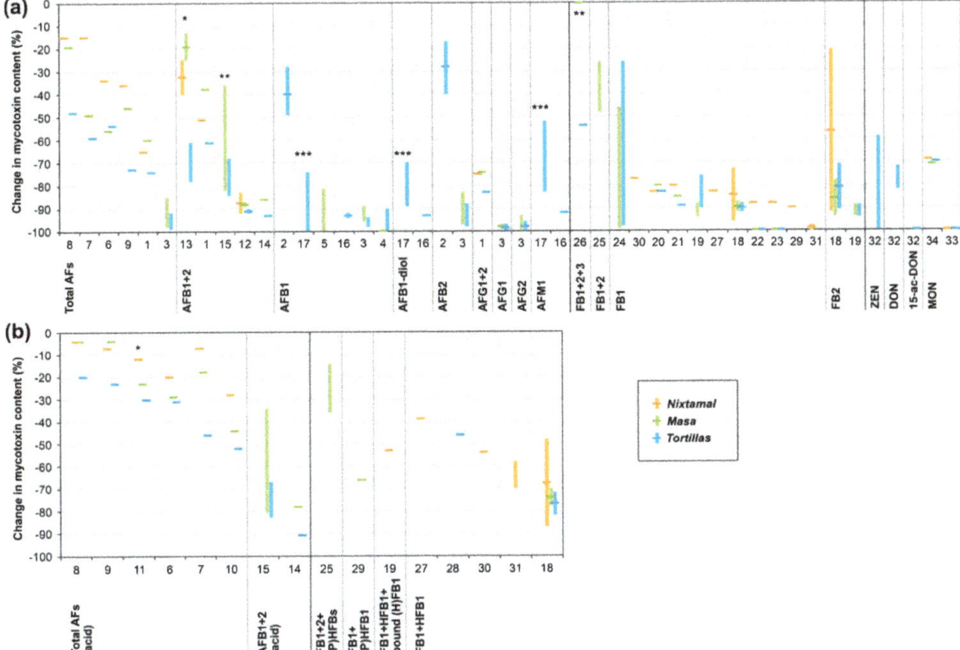

Figure 2. Graphical overview of mycotoxin changes during nixtamalization and tortilla production. (**a**) Free parent mycotoxins. (**b**) Sum of free parent form(s) and detected modified/ matrix-associated form(s). Columns indicate (approximate) ranges; lines represent (approximate) mean values of changes in mycotoxin concentrations from raw maize material to tortillas or intermediate products. The numbers on the x-axis refer to studies described in the literature, which are listed in Tables 1–3. More than one number can refer to the same reference if different process conditions or technologies were compared. For details on the studies (including references, processing parameters, and mycotoxin content in raw maize material), see Tables 1–3. *: Only mixing of kernels with hot lime water, without further cooking. **: Microwave cooking. ***: Extrusion cooking. Total AFs: aflatoxins B1 + B2 + G1 + G2; Total AFs (acid): aflatoxins B1 + B2 + G1 + G2 detected in acidified samples/extracts; AFB1/2: aflatoxin B1/2; AFB1 + 2 (acid): aflatoxin B1 + B2 detected in acidified samples/extracts; AFB1-diol: aflatoxin B1 dihydrodiol; AFG1/2: aflatoxin G1/2; AFM1: aflatoxin M1; FB1/2/3: fumonisin B1/2/3; (P)HFBs: partly + fully hydrolyzed fumonisins B1 + B2; (P)HFB1: partly + fully hydrolyzed fumonisin B1; ZEN: zearalenone; DON: deoxynivalenol; 15-ac-DON: 15-acetyl-deoxynivalenol; MON: moniliformin.

Funding: This work was performed as part of the MyToolBox project (www.mytoolbox.eu), which has received funding from the European Union's Horizon 2020 research and innovation program under grant agreement no. 678012. The article reflects only the authors' view, and the European Commission is not responsible for any use that may be made of the information it contains.

Acknowledgments: We appreciate the work of all colleagues on this topic and apologize to those whose studies were not cited here.

Conflicts of Interest: The authors declare no conflicts of interest.

References

1.   Pitt, J.; Wild, C.; Baan, R.; Gelderblom, W.; Miller, J.; Riley, R.; Wu, F. *Improving Public Health through Mycotoxin Control. Chapter 1: Fungi Producing Significant Mycotoxins*; International Agency for Research on Cancer, WHO Press: Geneva, Switzerland, 2012; pp. 1–30.

2. Taniwaki, M.H.; Pitt, J.I.; Magan, N. *Aspergillus* species and mycotoxins: Occurrence and importance in major food commodities. *Curr. Opin. Food Sci.* **2018**, *23*, 38–43. [CrossRef]

3. EFSA CONTAM. Opinion of the scientific panel on contaminants in the food chain [CONTAM] related to the potential increase of consumer health risk by a possible increase of the existing maximum levels for aflatoxins in almonds, hazelnuts and pistachios and derived products. *EFSA J.* **2007**, *446*, 1–127.

4. EFSA Panel on Contaminants in the Food Chain; Knutsen, H.-K.; Barregård, L.; Bignami, M.; Brüschweiler, B.; Ceccatelli, S.; Cottrill, B.; Dinovi, M.; Edler, L.; Grasl-Kraupp, B.; et al. Appropriateness to set a group health-based guidance value for fumonisins and their modified forms. *EFSA J.* **2018**, *16*, e05172.

5. Metzler, M. Proposal for a uniform designation of zearalenone and its metabolites. *Mycotoxin Res.* **2011**, *27*, 1–3. [CrossRef] [PubMed]

6. EFSA CONTAM. Scientific Opinion on the risks for public health related to the presence of zearalenone in food. *EFSA J.* **2011**, *9*, 2197. [CrossRef]

7. EFSA. Deoxynivalenol in food and feed: Occurrence and exposure. *EFSA J.* **2013**, *11*, 3379.

8. Jestoi, M. Emerging *Fusarium*-mycotoxins fusaproliferin, beauvericin, enniatins, and moniliformin—A review. *Crit. Rev. Food Sci. Nutr.* **2008**, *48*, 21–49. [CrossRef] [PubMed]

9. Kovač, M.; Šubarić, D.; Bulaić, M.; Kovač, T.; Šarkanj, B. Yesterday *masked*, today *modified*; what do mycotoxins bring next? *Arh. Hig. Rada Toksikol.* **2018**, *69*, 196–214. [CrossRef] [PubMed]

10. Dall'Asta, C.; Berthiller, F. *Masked Mycotoxins in Food: Formation, Occurrence and Toxicological Relevance*; Royal Society of Chemistry: Cambridge, UK, 2016.

11. Rychlik, M.; Humpf, H.-U.; Marko, D.; Dänicke, S.; Mally, A.; Berthiller, F.; Klaffke, H.; Lorenz, N. Proposal of a comprehensive definition of modified and other forms of mycotoxins including "masked" mycotoxins. *Mycotoxin Res.* **2014**, *30*, 197–205. [CrossRef] [PubMed]

12. Rooney, L.W.; Serna-Saldivar, S.O. Tortillas. In *Reference Module in Food Science*; Elsevier: Amsterdam, The Netherlands, 2016. [CrossRef]

13. Serna-Saldivar, S.O. Chapter 1—History of corn and wheat tortillas. In *Tortillas*; Rooney, L.W., Serna-Saldivar, S.O., Eds.; AACC International Press: St. Paul, MN, USA, 2015; pp. 1–28.

14. Santiago-Ramos, D.; Figueroa-Cárdenas, J.d.D.; Mariscal-Moreno, R.M.; Escalante-Aburto, A.; Ponce-García, N.; Véles-Medina, J.J. Physical and chemical changes undergone by pericarp and endosperm during corn nixtamalization—A review. *J. Cereal Sci.* **2018**, *81*, 108–117. [CrossRef]

15. Serna-Saldivar, S.O. Chapter 2—Nutrition and fortification of corn and wheat tortillas. In *Tortillas*; Rooney, L.W., Serna-Saldivar, S.O., Eds.; AACC International Press: St. Paul, MN, USA, 2015; pp. 29–63.

16. Bressani, R.; Scrimshaw, N.S. Lime-heat effects on corn nutrients, effect of lime treatment on in vitro availability of essential amino acids and solubility of protein fractions in corn. *J. Agric. Food Chem.* **1958**, *6*, 774–778. [CrossRef]

17. Rendon-Villalobos, R.; Bello-Pérez, L.A.; Osorio-Díaz, P.; Tovar, J.; Paredes-López, O. Effect of storage time on in vitro digestibility and resistant starch content of nixtamal, masa, and tortilla. *Cereal Chem.* **2002**, *79*, 340–344. [CrossRef]

18. Mariscal-Moreno, R.M.; de Dios Figueroa Cárdenas, J.; Santiago-Ramos, D.; Rayas-Duarte, P.; Veles-Medina, J.J.; Martínez-Flores, H.E. Nixtamalization process affects resistant starch formation and glycemic index of tamales. *J. Food Sci.* **2017**, *82*, 1110–1115. [CrossRef]

19. Pilcher, J. Taste, Smell, and Flavor in Mexico. In *Oxford Research Encyclopedia of Latin American History*; Oxford University Press: New York, NY, USA, 2016. Available online: http://oxfordre.com/latinamericanhistory/ (accessed on 5 April 2019).

20. de Arriola, M.d.C.; de Porres, E.; de Cabrera, S.; de Zepeda, M.; Rolz, C. Aflatoxin fate during alkaline cooking of corn for tortilla preparation. *J. Agric. Food Chem.* **1988**, *36*, 530–533. [CrossRef]

21. Ulloa-Sosa, M.; Schroeder, H.W. Note on aflatoxin decomposition in the process of making tortillas from corn. *Cereal Chem.* **1969**, *46*, 397–400.

22. Abbas, H.K.; Mirocha, C.J.; Rosiles, R.; Carvajal, M. Effect of tortilla-preparation process on aflatoxins B1 and B2 in corn. *Mycotoxin Res.* **1988**, *4*, 33–36. [CrossRef] [PubMed]

23. Price, R.L.; Jorgensen, K.V. Effects of processing on aflatoxin levels and on mutagenic potential of tortillas made from naturally contaminated corn. *J. Food Sci.* **1985**, *50*, 347–349. [CrossRef]

24. Moreno-Pedraza, A.; Valdés-Santiago, L.; Hernández-Valadez, L.J.; Rodríguez-Sixtos Higuera, A.; Winkler, R.; Guzmán-de Peña, D.L. Reduction of aflatoxin B1 during tortilla production and identification of degradation by-products by direct-injection electrospray mass spectrometry. *Salud Pública de México* **2015**, *57*, 50–56. [CrossRef]

25. Guzman-de-Pena, D.; Trudel, L.; Wogan, G.N. Corn "nixtamalización" and the fate of radiolabelled aflatoxin B1 in the tortilla making process. *Bull. Environ. Contam. Toxicol.* **1995**, *55*, 858–864. [CrossRef]

26. Pérez-Flores, G.C.; Moreno-Martínez, E.; Méndez-Albores, A. Effect of microwave heating during alkaline-cooking of aflatoxin contaminated maize. *J. Food Sci.* **2011**, *76*, T48–T52. [CrossRef]

27. Rosentrater, K.A. A review of corn masa processing residues: Generation, properties, and potential utilization. *Waste Manag.* **2006**, *26*, 284–292. [CrossRef]

28. Bazúa, C.D.; Guerra, R.; Sterner, H. Extruded corn flour as an alternative to lime-heated corn flour for tortilla preparation. *J. Food Sci.* **1979**, *44*, 940–941. [CrossRef]

29. Gómez-Aldapa, C.; Martínez-Bustos, F.; Figueroa, C.J.D.; Ordorica, F.C.A. A comparison of the quality of whole corn tortillas made from instant corn flours by traditional or extrusion processing. *Int. J. Food Sci. Technol.* **1999**, *34*, 391–399. [CrossRef]

30. Elias-Orozco, R.; Castellanos-Nava, A.; Gaytán-Martínez, M.; Figueroa-Cárdenas, J.D.; Loarca-Piña, G. Comparison of nixtamalization and extrusion processes for a reduction in aflatoxin content. *Food Addit. Contam.* **2002**, *19*, 878–885. [CrossRef]

31. Lutz, W.K.; Jaggi, W.; Lüthy, J.; Sagelsdorff, P.; Schlatter, C. In vivo covalent binding of aflatoxin B1 and aflatoxin M1 to liver DNA of rat, mouse and pig. *Chem.-Biol. Interact.* **1980**, *32*, 249–256. [CrossRef]

32. Pallavi, R.M.V.; Vidyasagar, T.; Rao, B. Production of aflatoxin M1 by *Aspergillus parasiticus* (AP 456) in a semisynthetic medium. *Indian J. Exp. Biol.* **1997**, *35*, 735–741.

33. Pai, M.R.; Bai, N.J.; Venkitasubramanian, T.A. Production of aflatoxin M in a liquid medium. *Appl. Microbiol.* **1975**, *29*, 850–851. [PubMed]

34. Vesonder, R.; Haliburton, J.; Stubblefield, R.; Gilmore, W.; Peterson, S. *Aspergillus flavus* and aflatoxins B_1, B_2, and M_1 in corn associated with equine death. *Arch. Environ. Contam. Toxicol.* **1991**, *20*, 151–153. [CrossRef] [PubMed]

35. Méndez-Albores, J.A.; Arámbula-Villa, G.; Loarca-Piña, M.G.; González-Hernández, J.; Castaño-Tostado, E.; Moreno-Martínez, E. Aflatoxins' fate during the nixtamalization of contaminated maize by two tortilla-making processes. *J. Stored Prod. Res.* **2004**, *40*, 87–94. [CrossRef]

36. Torres, P.; Guzmán-Ortiz, M.; Ramírez-Wong, B. Revising the role of pH and thermal treatments in aflatoxin content reduction during the tortilla and deep frying processes. *J. Agric. Food Chem.* **2001**, *49*, 2825–2829. [CrossRef]

37. Lee, L.S.; Dunn, J.J.; DeLucca, A.J.; Ciegler, A. Role of lactone ring of aflatoxin B1 in toxicity and mutagenicity. *Experientia* **1981**, *37*, 16–17. [CrossRef]

38. Vázquez-Durán, A.; Díaz-Torres, R.; Ramírez-Noguera, P.; Moreno-Martínez, E.; Méndez-Albores, A. Cytotoxic and genotoxic evaluation of tortillas produced by microwave heating during alkaline-cooking of aflatoxin-contaminated maize. *J. Food Sci.* **2014**, *79*, T1024–T1029. [CrossRef]

39. Méndez-Albores, J.; Villa, G.; Del Rio-García, J.; Martínez, E. Aflatoxin-detoxification achieved with Mexican traditional nixtamalization process (MTNP) is reversible. *J. Sci. Food Agric.* **2004**, *84*, 1611–1614. [CrossRef]

40. Beckwith, A.C.; Vesonder, R.F.; Ciegler, A. Action of weak bases upon aflatoxin B1 in contact with macromolecular reactants. *J. Agric. Food Chem.* **1975**, *23*, 582–587. [CrossRef]

41. Lee, L.S.; Conkerton, E.J.; Ory, R.L.; Bennett, J.W. [^{14}C]Aflatoxin B1 as an indicator of toxin destruction during ammoniation of contaminated peanut meal. *J. Agric. Food Chem.* **1979**, *27*, 598–602. [CrossRef]

42. Brekke, O.L.; Peplinski, A.J.; Lancaster, E.B. Aflatoxin inactivation in corn by aqua ammonia. *Trans. ASAE* **1977**, *20*, 1160–1168. [CrossRef]

43. Weng, C.Y.; Martinez, A.J.; Park, D.L. Efficacy and permanency of ammonia treatment in reducing aflatoxin levels in corn. *Food Addit. Contam.* **1994**, *11*, 649–658. [CrossRef]

44. Sydenham, E.W.; Stockenstrom, S.; Thiel, P.G.; Shephard, G.S.; Koch, K.R.; Marasas, W.F.O. Potential of alkaline hydrolysis for the removal of fumonisins from contaminated corn. *J. Agric. Food Chem.* **1995**, *43*, 1198–1201. [CrossRef]

45. Voss, K.A.; Poling, S.M.; Meredith, F.I.; Bacon, C.W.; Saunders, D.S. Fate of fumonisins during the production of fried tortilla chips. *J. Agric. Food Chem.* **2001**, *49*, 3120–3126. [CrossRef]

46. Dombrink-Kurtzman, M.A.; Dvorak, T.J.; Barron, M.E.; Rooney, L.W. Effect of nixtamalization (alkaline cooking) on fumonisin-contaminated corn for production of masa and tortillas. *J. Agric. Food Chem.* **2000**, *48*, 5781–5786. [CrossRef]

47. Palencia, E.; Torres, O.; Hagler, W.; Meredith, F.I.; Williams, L.D.; Riley, R.T. Total fumonisins are reduced in tortillas using the traditional nixtamalization method of mayan communities. *J. Nutr.* **2003**, *133*, 3200–3203. [CrossRef] [PubMed]

48. Méndez-Albores, A.; Cardenas-Rodriguez, D.A.; Vazquez-Duran, A. Efficacy of microwave-heating during alkaline processing of fumonisin-contaminated maize. *Iran J. Public Health* **2014**, *43*, 147–155.

49. Scudamore, K.; Scriven, F.; Patel, S. *Fusarium* mycotoxins in the food chain: Maize-based snack foods. *World Mycotoxin J.* **2009**, *2*, 441–450. [CrossRef]

50. De La Campa, R.; Miller, J.D.; Hendricks, K. Fumonisin in tortillas produced in small-scale facilities and effect of traditional masa production methods on this mycotoxin. *J. Agric. Food Chem.* **2004**, *52*, 4432–4437. [CrossRef]

51. De Girolamo, A.; Lattanzio, V.M.T.; Schena, R.; Visconti, A.; Pascale, M. Effect of alkaline cooking of maize on the content of fumonisins B1 and B2 and their hydrolysed forms. *Food Chem.* **2016**, *192*, 1083–1089. [CrossRef]

52. Gelderblom, W.C.A.; Cawood, M.E.; Snyman, S.D.; Vleggaar, R.; Marasas, W.F.O. Structure-activity relationships of fumonisins in short-term carcinogenesis and cytotoxicity assays. *Food Chem. Toxicol.* **1993**, *31*, 407–414. [CrossRef]

53. Grenier, B.; Bracarense, A.-P.F.L.; Schwartz, H.E.; Trumel, C.; Cossalter, A.-M.; Schatzmayr, G.; Kolf-Clauw, M.; Moll, W.-D.; Oswald, I.P. The low intestinal and hepatic toxicity of hydrolyzed fumonisin B1 correlates with its inability to alter the metabolism of sphingolipids. *Biochem. Pharmacol.* **2012**, *83*, 1465–1473. [CrossRef]

54. Voss, K.A.; Riley, R.T.; Snook, M.E.; Waes, J.G.-V. Reproductive and sphingolipid metabolic effects of fumonisin B1 and its alkaline hydrolysis product in LM/Bc mice: Hydrolyzed fumonisin B1 did not cause neural tube defects. *Toxicol. Sci.* **2009**, *112*, 459–467. [CrossRef] [PubMed]

55. Park, J.W.; Scott, P.M.; Lau, B.P.-Y.; Lewis, D.A. Analysis of heat-processed corn foods for fumonisins and bound fumonisins. *Food Addit. Contam.* **2004**, *21*, 1168–1178. [CrossRef]

56. Burns, T.D. Fate of the Mycotoxin Fumonisin B1 During Alkaline Cooking of Cultured and Whole Kernel Corn. Ph.D. Thesis, University of Georgia, Georgia, Athens, 2008.

57. Cortez-Rocha, M.O.; Trigo-Stockli, D.M.; Wetzel, D.L.; Reed, C.R. Effect of extrusion processing on fumonisin B1 and hydrolyzed fumonisin B1 in contaminated alkali-cooked corn. *Bull. Environ. Contam. Toxicol.* **2002**, *69*, 471–478. [CrossRef] [PubMed]

58. Howard, P.C.; Churchwell, M.I.; Couch, L.H.; Marques, M.M.; Doerge, D.R. Formation of N-(carboxymethyl) fumonisin B1, following the reaction of fumonisin B1 with reducing sugars. *J. Agric. Food Chem.* **1998**, *46*, 3546–3557. [CrossRef]

59. Seefelder, W.; Knecht, A.; Humpf, H.-U. Bound fumonisin B1: Analysis of fumonisin-B1 glyco and amino acid conjugates by liquid chromatography–electrospray ionization–tandem mass spectrometry. *J. Agric. Food Chem.* **2003**, *51*, 5567–5573. [CrossRef] [PubMed]

60. Voss, K.A.; Riley, R.T.; Jackson, L.S.; Jablonski, J.E.; Bianchini, A.; Bullerman, L.B.; Hanna, M.A.; Ryu, D. Extrusion cooking with glucose supplementation of fumonisin-contaminated corn grits protects against nephrotoxicity and disrupted sphingolipid metabolism in rats. *Mol. Nutr. Food Res.* **2011**, *55*, S312–S320. [CrossRef]

61. Park, J.W.; Scott, P.M.; Lau, B.P.Y. Analysis of N-fatty acyl fumonisins in alkali-processed corn foods. *Food Sci. Biotechnol.* **2013**, *22*, 147–152. [CrossRef]

62. Burns, T.D.; Snook, M.E.; Riley, R.T.; Voss, K.A. Fumonisin concentrations and in vivo toxicity of nixtamalized *Fusarium verticillioides* culture material: Evidence for fumonisin–matrix interactions. *Food Chem. Toxicol.* **2008**, *46*, 2841–2848. [CrossRef]

63. Voss, K.A.; Riley, R.T.; Moore, N.D.; Burns, T.D. Alkaline cooking (nixtamalisation) and the reduction in the in vivo toxicity of fumonisin-contaminated corn in a rat feeding bioassay. *Food Addit. Contam. Part A* **2013**, *30*, 1415–1421. [CrossRef] [PubMed]

64. Hendrich, S.; Miller, K.A.; Wilson, T.M.; Murphy, P.A. Toxicity of *Fusarium proliferatum*-fermented nixtamalized corn-based diets fed to rats: Effect of nutritional status. *J. Agric. Food Chem.* **1993**, *41*, 1649–1654. [CrossRef]

65. Abbas, H.K.; Mirocha, C.J.; Rosiles, R.; Carvajal, M. Decomposition of zearalenone and deoxynivalenol in the process of making tortillas from corn. *Cereal Chem.* **1988**, *65*, 15–19.

66. Trenholm, H.L.; Charmley, L.L.; Prelusky, D.B.; Warner, R.M. Washing procedures using water or sodium carbonate solutions for the decontamination of three cereals contaminated with deoxynivalenol and zearalenone. *J. Agric. Food Chem.* **1992**, *40*, 2147–2151. [CrossRef]
67. Schaarschmidt, S.; Fauhl-Hassek, C. The fate of mycotoxins during the processing of wheat for human consumption. *Compr. Rev. Food Sci. Food Saf.* **2018**, *17*, 556–593. [CrossRef]
68. Pineda-Valdes, G.; Ryu, D.; Jackson, D.S.; Bullerman, L.B. Reduction of moniliformin during alkaline cooking of corn. *Cereal Chem.* **2002**, *79*, 779–782. [CrossRef]
69. Pineda-Valdes, G.; Bullerman, L.B. Thermal stability of moniliformin at varying temperature, pH, and time in an aqueous environment. *J. Food Prot.* **2000**, *63*, 1598–1601. [CrossRef] [PubMed]

MDPI
St. Alban-Anlage 66
4052 Basel
Switzerland
Tel. +41 61 683 77 34
Fax +41 61 302 89 18
www.mdpi.com

Toxins Editorial Office
E-mail: toxins@mdpi.com
www.mdpi.com/journal/toxins